Petroleum Production
in
Nontechnical Language
SECOND EDITION

Petroleum Production
in
Nontechnical Language
SECOND EDITION

FOREST GRAY

PennWell Books

PENNWELL PUBLISHING COMPANY
TULSA, OKLAHOMA

Copyright © 1995 by
PennWell Publishing Company
1421 South Sheridan/P.O. Box 1260
Tulsa, Oklahoma 74101

Library of Congress Cataloging-in-Publication Data
Gray, Forest
   Petroleum production in nontechnical language / Forest Gray. – 2nd ed.
   Includes index.
   ISBN 0-87814-450-1
   1. Petroleum engineering.   I. Gray, Forest.   Petroleum production for the
nontechnical person.   II.   Title.
TN870 . G732   1995
665.5–dc20

Printed in the United States of America

1 2   3   4   5   99   98   97   96   95

Contents

Foreword

For as long as I have been associated with the oil industry, which is only about 36 years, I have been aware of the lack of tools available to the non-engineers to permit self-improvement. It's as if there is a technocratic conspiracy to keep the technology in the technicians. Or maybe there is a caste system in oil companies. Those with the engineering degrees don't need a primer on technology; those without the technical degrees don't have a need to know, so no technician attempted to articulate the fundamentals.

My experience is that both situations probably exist from time to time and in varying amounts, so I take pen and ruler in hand to combat the first and disprove the second. This book is intended to satisfy the needs of the young, astute, aspiring industry entrant who has no degree in petroleum engineering or petroleum geology or one of the closely related disciplines. It will be useful to many persons in financial work, supplies, transportation, distribution, public relations, advertising, sales, or purchasing in petroleum based or related companies.

The format is designed to enlighten the reader a little bit at a time, followed by some reinforcing exercises which often build on previous work. If nothing else, by the time the reader is finished with the book he probably will not be intimidated by technical jargon. At least he will be able to ask more penetrating questions. And that's most of the self-improvement battle won.

Much of the material on which the book is based was taken from literature published by the Society of Petroleum Engineers of the American Institute of Mining, Metallurgical and Petroleum Engineers, the Division of Production of the American Petroleum Institute, the Natural Gasoline Association of America, and the American Association of Petroleum Geologists; specifically, L.C. Uren, *Oil Field Development*, McGraw-Hill, 1946; Thomas C. Frick et al., *Petroleum Production Engineering*, Vols. I and II, Society of Petroleum Engineers, 1962; and the U.S. Department of Energy, National Petroleum Council. In addition, the author is indebted to the host of authors who have contributed to the petroleum literature in various other publications.

CHAPTER I

Overview of the Petroleum Industry

The word *petroleum*—from the Latin *petra*, rock and *oleum*, oil—is properly applied to liquid hydrocarbons and is a perfectly good synonym for crude oil. It is also widely used to refer to natural gas. The term "petroleum industry" generally refers to both the oil and the gas industries and is used in that sense throughout this book. Thus construed, petroleum comprises a large and complex group of liquid, gaseous, and semisolid hydrocarbons, i.e., hydrogen/carbon compounds that often contain "impurities" such as nitrogen, oxygen, and sulfur.

The complex hydrocarbons that make up oil, natural gas, and coal are thought to be derived from dead animal and plant life laid down on ancient seafloors. Later, this organic matter was transformed through the heat and pressure generated by subsequent layers of sedimentary materials (see Chapter 3). In geological time, the initial deposition of this organic material was a comparatively recent event. All petroleum deposits discovered thus far originated within the last 10% of the earth's life (from 2 to 400 million years ago) during the Paleozoic, Mesozoic, and Tertiary Cenozoic eras.

With this brief technical description of petroleum behind us, let's now look at the historical and economic aspects of the oil industry.

STAGES OF HISTORICAL DEVELOPMENT

In 1859, two oil wells in the United States, with an estimated combined value of $40,000, produced 2,000 barrels (bbl) of crude oil. More than a century later, hundreds of thousands of wells were in operation in the U.S., with an annual crude oil production of more than 2.3 billion bbl and an estimated wellhead value of more than $35 billion.

Between these two points in time, of course, many things happened in and to the petroleum industry.

Table 1-1
Stages of Historical Development for the Petroleum Industry

Stage 1   **Wild exploitation (1859-1875)**
Stage 2   **Domination of Standard Oil Group (1870-1910)**
Stage 3   **Modern Era; dissolution of Standard Oil Group empire and emergence of new companies (1911-1928)**
Stage 4   **Great Depression of the Thirties and government regulation of production for the first time (1930-1945)**
Stage 5   **Period of competitive realignment and worldwide growth; rapid development of natural gas industry (1945-present)**

The historical development of the U.S. petroleum industry consists of five distinct stages, described below.

I. A period of wild exploitation began in 1859 when "Colonel" Edwin L. Drake drilled the first commercially successful oil well near Titusville, Pennsylvania. This period, which extended well into the 1870s, was a time of frantic competition, an economic free-for-all centering around production and characterized by a combination of energy, avarice, and ignorance reminiscent of the California Gold Rush days. A highly erratic price structure afforded easy wealth to some and quick ruin to others. Thus, through the successful exploitation of the oil-producing potential of the eastern states, the foundations of a great industry were laid.

II. For 40 years, beginning in 1870 with the formation of the Standard Oil Company (Ohio), the entire activity of the petroleum industry was dominated by the Standard Oil group. During this period, the Standard Oil group controlled many of the refining, transportation, and marketing functions of the industry, thereby assuring price control of the fourth principal function, production. By the early 1880s the Standard Oil group controlled 80-90% of all U.S. refineries. By 1904, the Standard distribution system—a vast network of railroads, pipelines, bulk plants, and affiliated marketing outlets—extended into most states. Close to 90% of all U.S. petroleum dealers at that time bought from the Standard Oil group.

The growth and rise to power of the Standard Oil group began by horizontal integration at the refinery level and later developed into vertical integration through a succession of affiliations with railroads and pipelines; with producers, oil buyers, and refiners in other areas; with oil gathering and trunk lines; and with various other types of regional and local marketing organizations.

While the Standard group was using vertical integration to monopolize, other industry elements began to use it as a defensive mechanism in the struggle for competitive survival. Thus integration, which remains today an

Fig. 1-1
Wooden derricks like this were common early in the Twentieth Century
(courtesy Kern County Museum).

important part of the petroleum industry's way of life, has developed historically largely as a result of the 40-year dominance of the Standard Oil Trust.

III. The dissolution, between 1911 and 1915, of the Standard empire
restored competition, resulting not only in the emergence of many new
companies but in a continuation of the search for security, through
integration, by the "major" oil companies. This third istorical stage is
sometimes called the modern era of the petroleum industry.

As competition was restored and oil demand increased during World War
I, the petroleum industry expanded to meet the burgeoning requirements of
the U.S. industry and all the newly developed means of land, sea, and air
transportation that depended upon petroleum products for fuel and lubrication. A period of remarkable activity resulted and an enormous and complex
industry developed, ranging from massive integrated firms operating
worldwide (including the heirs of the old Standard Oil organization) to
single-individual enterprises specializing in one facet of exploration,
production, or retail distribution.

IV. The 1930s brought the Great Depression and government regulation
of production, which has continued ever since as a significant political and economic fact of life for the petroleum industry.

The petroleum industry's chief troubles usually have arisen from overpro-
duction: too much output, too little demand, too low a price, too little profit.
Those few attempts made at voluntarily restricting crude oil output to restore
a profitable price structure almost invariably collapsed as crude oil producers
(including royalty owners) succumbed to the pressure of economic necessity
or the enticement of profits. Another factor contributing to the failure of most
such efforts at self-regulation was the recurring, disturbing suspicion that we
were fast approaching the "bottom of the barrel" of our crude oil reserves, in
which case, obviously, the smart operator should hunt frantically and develop
feverishly while there was still time. Even some of the largest companies found
it difficult to ignore this suspicion and to refrain from acting like smart operators. The consequences of these practices were a series of federal and state
interventions, urgently requested by the petroleum industry. These became
effective just as a new war loomed, with the prospect of virtually limitless
demand.

This was also the period in which liquefied petroleum gas (LPG) and natural gasoline, previously considered field waste products, started to come into
their own, for former largely as heating fuel and the latter as a blending agent
for the new "composite" gasolines.

V. After World War II, a period of competitive realignment in the late
1940s and early 1950s found some U.S. companies with substantial foreign

Fig. 1-2
Spindletop, near Beaumont, Tex., was the famed gusher that heralded the blossoming of the modern petroleum era.

Fig. 1-3
The early years of the petroleum industry were marked by boom and bust cycles. At the turn of the century, California experienced a second gold rush–one for "black gold"–that drew a rugged breed of risk takers, such as these shown in Kern River oil field ca. 1908-1910 (from the collection of the Petroleum Production Pioneers).

crude oil resources and others almost exclusively dependent on domestic production. In this period, the U.S. petroleum industry, slightly tinged with an international flavor almost from its inception, became truly worldwide in its relationships and correspondingly more sensitive to international political and economic developments.

In this same stage the natural gas industry developed remarkably. Natural gas trunk lines, often 2,000 miles or more in length, began to move enormous quantities of a standardized product to consuming centers throughout the nation. Today natural gas ranks as one of the 10 largest U.S. industries in terms of capital investment and provides more than 25% of the nation's total energy requirements. Oil had barely established itself as a successful competitor of coal in the heating markets when it felt that rival impact of natural gas.

Although the U.S. chemical industry had been growing rapidly since World War I—earlier it had been limited largely to the long-existent explosives business—it was not until after World War II that crude oil and natural gas became primary raw material sources for the new plastics and synthetics. Thus developed one of America's newest industries, petrochemicals—both a market for and an integral part of the petroleum business. Now more than one-fourth of the entire output of the U.S. chemical industry comes from petroleum.

No major project is undertaken without considering the economics involved. Drilling an oil well is indeed a major project, as the next section explains.

SOME ECONOMIC ASPECTS

The folklore of the petroleum industry has it that more money has gone down holes than ever came back up. Difficult as it might be to prove that statistically, the 80% dry-hole figure suggests it's right. One of the distinguishing characteristics of the U.S. petroleum industry in recent years has been an increasingly poor cost/profit relationship. This is largely because of rapidly mounting exploration and development costs.

Colonel Drake's discovery well was dug to 59 1/2 ft, in contrast to current (exceptional) well depths of 20,000 ft and more (The deepest Free World exploratory well to date was drilled in Washita County, Oklahoma, in 1974 to a depth of 31,441 ft. The maximum producing depth on record is 25,446 ft, a Pecos County, Texas, well completed in 1984). Entirely aside from the special technical problems encountered in drilling to such depths, costs seem to increase geometrically as wells are drilled to 15,000 ft and more.

During the boom days of the 1970s and early 1980s, the cost of oil field equipment shot up by more than 50%, and the costs of contract exploration services doubled in some areas over the same period. The cost of drilling a typical 10,000-ft wildcat (exploratory) well today may be anywhere from $50,000 to tens of millions of dollars, depending on drilling conditions, and the odds are that it will be a dry hole!

Partially mitigating this dim economic prospect in recent years have been (1) the discovery and development of substantial oil deposits in new areas of the world that each year emerge as highly prospective for hydrocarbons (2) the successful exploration for and technical mastery of deepwater offshore and undersea oil production as well as in arctic and other remote, often severe environment areas. Although arctic and deepwater offshore drilling methods are by far the most costly type of drilling yet attempted, the enormous potential of the new oil and gas reserves may justify the investment risk for most companies.

Today the estimated total of proved crude oil reserves of the Free World approaches 1 trillion bbl. Of this supply, more than 75% is in countries belonging to the Organization of Petroleum Exporting Countries (OPEC), and about two-thirds of the world total lies within the Middle East. The Western Hemisphere currently accounts for only 15% of the world's total proved crude oil reserves but produces about 25% of the world's crude output of about 69 million barrels per day (b/d) from about 70% of the world's oil wells.

Only a decade ago, in terms of daily output, crude oil production from the Western Hemisphere outstripped that of the Middle East by about 50%.

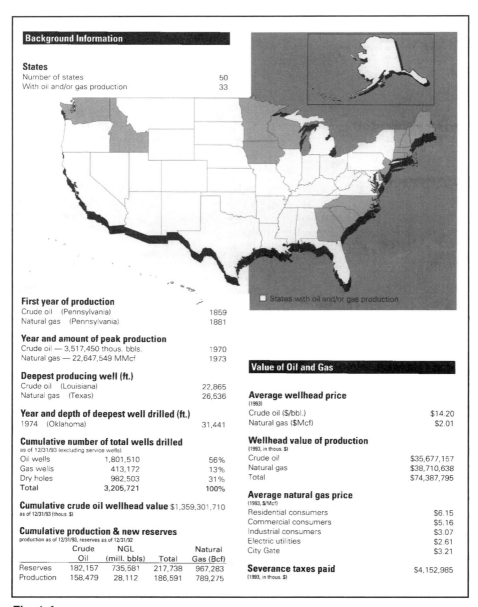

Background Information

States

Number of states	50
With oil and/or gas production	33

☐ States with oil and/or gas production

First year of production

Crude oil (Pennsylvania)	1859
Natural gas (Pennsylvania)	1881

Year and amount of peak production

Crude oil — 3,517,450 thous. bbls.	1970
Natural gas — 22,647,549 MMcf	1973

Deepest producing well (ft.)

Crude oil (Louisiana)	22,865
Natural gas (Texas)	26,536

Year and depth of deepest well drilled (ft.)

1974 (Oklahoma)	31,441

Cumulative number of total wells drilled
as of 12/31/93 (excluding service wells)

Oil wells	1,801,510	56%
Gas wells	413,172	13%
Dry holes	982,503	31%
Total	3,205,721	100%

Cumulative crude oil wellhead value $1,359,301,710
as of 12/31/93 (thous. $)

Cumulative production & new reserves
production as of 12/31/93, reserves as of 12/31/92

	Crude Oil	NGL (mill. bbls)	Total	Natural Gas (Bcf)
Reserves	182,157	735,581	217,738	967,283
Production	158,479	28,112	186,591	789,275

Value of Oil and Gas

Average wellhead price
(1993)

Crude oil ($/bbl.)	$14.20
Natural gas ($Mcf)	$2.01

Wellhead value of production
(1993, in thous. $)

Crude oil	$35,677,157
Natural gas	$38,710,638
Total	$74,387,795

Average natural gas price
(1993, $/Mcf)

Residential consumers	$6.15
Commercial consumers	$5.16
Industrial consumers	$3.07
Electric utilities	$2.61
City Gate	$3.21

Severance taxes paid
(1993, in thous. $) $4,152,985

Fig. 1-4
A snapshot of the U.S. oil and gas producing industry today (courtesy of the Independent Petroleum Association of America).

1993 Industry Statistics

Number of wells drilled

	Exploratory	Development	Total
Oil	460	7,759	8,219
Gas	438	8,096	8,534
Dry	2,397	3,965	6,362
Service	–	702	702
Total	3,295	20,522	23,817

Total footage drilled
(thous. ft.)

	Exploratory	Development	Total
Oil	2,883.5	40,904.5	43,788.0
Gas	2,645.5	50,566.0	53,211.5
Dry	13,635.0	19,530.6	33,165.6
Service	–	2,365.4	2,365.4
Total	19,164.0	113,366.5	132,530.5

(Note: Totals may not add due to rounding.)

New-field wildcats drilled

	1,567
Footage (thous. Ft.)	9,636.7

Average rotary rigs active

755

World-wide rank

	Crude Oil	Natural Gas
Wells drilled	1st	1st
Production	3rd	2nd
Reserves (1992)	10th	6th

Number of producing wells
(12/31/93)

Crude oil	583,879
Flowing	34,627
Artificial lift	549,252
Natural gas	286,168
Total	870,047

Average production

	thous. bbls.	thous. b/d
Crude oil	2,499,033	7,191
NGL (est.)	619,405	1,697
Total	3,244,037	8,888

Natural gas marketed production
(MMcf)

19,240,056

Average output per producing well

Crude oil (bbls.)	4,315
Natural gas (Mcf)	68,084

Average number of employees

Oil and natural gas extraction	340,390
Refining	124,014
Transportation	176,569
Wholesale	163,265
Retail	607,963
Total petroleum industry	1,412,201

1992 Latest Available Data

Petroleum reserves
as of 12/31/92 (mill. bbls.)

	Crude Oil	NGL	Total
New reserves	1,509	760	2,269
Production	2,446	773	3,219
Net annual change	-937	-13	-950
Proved reserves	23,745	7,451	31,194

Natural gas reserves
as of 12/31/92 (Bcf)

	Associated Dissolved	Non-Associated	Dry Gas
New reserves	2,638	13,615	15,376
Production	3,031	15,238	17,423
Net annual change	-393	-1,623	-2,047
Proved reserves	31,424	141,885	165,015

Cost of drilling and equipping wells

	Cost/ft. ($)	Cost/ well ($)	Total Cost (thous. $)
Oil	69.51	362,260	3,123,041
Gas	72.83	426,114	3,231,222
Dry	67.82	357,624	2,211,545
Total	70.27	332,607	8,565,808

Stripper wells

Producing stripper wells	453,277
Stripper well abandonments	16,211
Crude oil production in bbls.	368,131,720
Crude oil production b/d	1,008,580
Percentage of oil production	15%

Stripper oil reserves
as of 12/31/92 (thous. bbls.)

Primary	1,709,320
Secondary	1,562,210
Total	3,271,530

Federal Onshore Mineral Lease Royalties

Oil	$280,355,318
Gas	$243,635,393
Total Royalties	$865,437,216

Federal Onshore Oil and Gas Leases

Number of leases	19,428
Acres leased	10,710,890

For more information please contact: Information Services Department, Independent Petroleum Association of America, 1101 16th Street, N.W., Washington, D.C. 20036, 202-857-4760, FAX: 202-857-4799

Fig. 1-4
(continued)

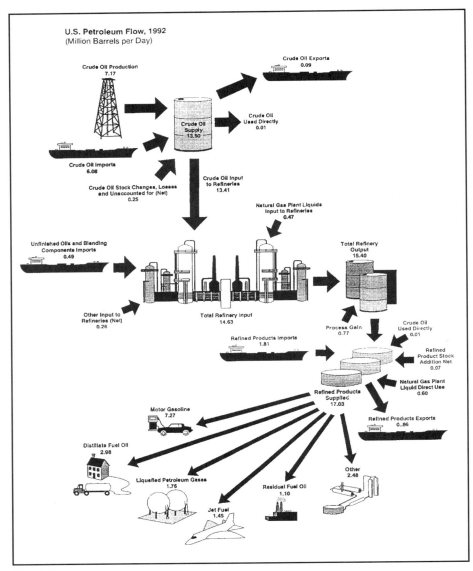

Fig. 1-5
U.S. petroleum flow, 1992. Source: Energy Information Administration (EIA).

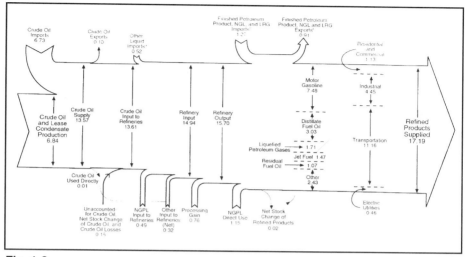

Fig. 1-6
U.S. petroleum flow, 1993. Source: EIA.

However, today, the Middle East outproduces the Western Hemisphere, and that trend is accelerating.

Little real rivalry exists between the two areas as far as the economics of oil production are concerned: Cost per barrel is so much less and output per well so much greater in the Middle East than in the Western Hemisphere that there is virtually no competition. Today, the United States has 100 times as many producing oil wells as does the Middle East but with only about one-third of the production. To cite a single example, fewer than 400 wells in Kuwait have been known to produce nearly one-fourth as much crude oil per day as the total daily output of hundreds of thousands of producing wells in the United States. The comparatively small output per well in the United States is due in part to the age of our production industry and the consequent large number of "stripper" operations. The trend of declining well productivity in the U.S. has persisted for decades and isn't likely to turn around to any significant degree anytime soon. By 1993, U.S. average oil well productivity had plunged to 12 b/d per well from peak productivity of more than 18 b/d/well achieved in the early 1970s. In 1994, U.S. annual average crude oil production had fallen to its lowest level in 40 years.

Table 1-2
Worldwide Crude Oil and Gas Production

1,000 b/d	Jan. 1995	Dec. 1994	1 month average Production 1995	1 month average Production 1994	Chg. vs prev. year Volume	Chg. vs prev. year %	Gas (bcf) Jan. 1995	Gas (bcf) Dec. 1994	Gas (bcf) Cum. 1995
Argentina	700	699	700	635	65	10.2	52.0	52.3	52.0
Bolivia	26	26	26	25	1	4.0	8.0	8.3	8.0
Brazil	708	709	708	658	50	7.6	8.7	10.1	8.7
Canada	1,781	1,792	1,781	1,671	110	6.6	564.9	550.2	564.9
Chile	11	12	11	14	−3	−21.4	3.4	3.4	3.4
Colombia	501	545	501	455	46	10.1	13.3	13.3	13.3
Ecuador	410	398	410	360	50	13.9	0.3	0.3	0.3
Mexico	2,680	2,673	2,680	2,724	−44	−1.6	113.3	112.5	113.3
Peru	135	118	135	133	2	1.5	3.9	3.9	3.9
Trinidad	125	129	125	128	−3	−2.3	18.8	18.4	18.8
United States	6,616	6,686	6,616	6,777	−161	−2.4	1,699.0	1,689.0	1,699.0
Venezuela*	2,450	2,510	2,450	2,400	50	2.1	70.7	72.4	70.7
Other Latin America	37	36	37	34	3	8.8	0.1	0.1	0.1
Western Hemisphere	**16,180**	**16,333**	**16,180**	**16,014**	**166**	**1.0**	**2,556.4**	**2,534.2**	**2,556.4**
Austria	23	23	23	24	−1	−4.2	4.9	5.1	4.9
Denmark	185	184	185	197	−12	−6.1	18.8	18.2	18.8
France	51	55	51	58	−7	−12.1	11.7	11.4	11.7
Germany	56	58	56	58	−2	−3.4	71.1	71.5	71.1
Italy	85	88	85	81	4	4.9	65.0	65.1	65.0
Netherlands	70	76	70	80	−10	−12.5	370.0	325.4	370.0
Norway	2,683	2,811	2,683	2,501	182	7.3	97.0	95.8	97.0
Spain	18	18	18	17	1	5.9	2.1	2.1	2.1
Turkey	68	69	68	73	−5	−6.8	0.6	0.6	0.6
United Kingdom	2,603	2,730	2,603	2,364	239	10.1	298.0	286.6	298.0
Other Western Europe	9	10	9	10	−1	−10.0	8.0	7.9	8.0
Western Europe	**5,851**	**6,122**	**5,851**	**5,463**	**388**	**7.1**	**947.2**	**889.7**	**947.2**
Albania	12	12	12	12	—	—	0.4	0.4	0.4
C.I.S.	6,853	6,947	6,853	7,418	−565	−7.6	2,380.2	2,344.9	2,380.2
Croatia	40	40	40	40	—	—	5.7	5.7	5.7
Hungary	73	70	73	71	2	2.8	15.0	15.0	15.0
Romania	136	136	136	137	−1	−0.7	63.9	63.9	63.9
Serbia	24	24	24	24	—	—	2.3	2.3	2.3
Other Eastern Europe	16	22	16	21	−5	−23.8	18.4	18.4	18.4
Eastern Europe & C.I.S.	**7,154**	**7,251**	**7,154**	**7,723**	**−569**	**−7.4**	**2,485.9**	**2,450.6**	**2,485.9**
Algeria*	750	750	750	750	—	—	151.1	151.1	151.1
Angola	580	555	580	530	50	9.4	1.7	1.7	1.7
Cameroon	100	110	100	110	−10	−9.1	—	—	—
Congo	180	185	180	185	−5	−2.7	—	—	—
Egypt	890	890	890	900	−10	−1.1	25.1	25.1	25.1
Gabon*	340	340	340	300	40	13.3	0.3	0.3	0.3
Libya*	1,370	1,380	1,370	1,380	−10	−0.7	19.1	19.2	19.1
Nigeria*	1,850	1,900	1,850	1,920	−70	−3.6	13.5	13.9	13.5
Tunisia	90	92	90	89	1	1.1	1.0	1.0	1.0
Zaire	30	28	30	25	5	20.0	—	—	—
Other Africa	19	17	19	15	4	26.7	0.1	0.1	0.1
Africa	**6,199**	**6,247**	**6,199**	**6,204**	**−5**	**−0.1**	**211.9**	**212.4**	**211.9**
Bahrain	105	105	105	106	−1	−0.9	12.3	13.6	12.3
Iran*	3,500	3,700	3,500	3,590	−90	−2.5	81.4	86.0	81.4
Iraq*	600	600	600	500	100	20.0	8.6	8.6	8.6
Kuwait*†	2,000	2,000	2,000	2,000	—	—	13.8	13.9	13.8
Oman	843	830	843	769	74	9.6	11.3	11.1	11.3
Qatar*	420	420	420	410	10	2.4	21.4	21.4	21.4
Saudi Arabia*†	8,000	8,000	8,000	8,000	—	—	95.0	95.1	95.0
Syria	615	595	615	550	65	11.8	13.3	12.9	13.3
United Arab Emirates*	2,220	2,200	2,220	2,220	—	—	72.6	71.9	72.6
Yemen	340	350	340	336	4	1.2	—	—	—
Other Middle East	0	0	0	0	0	—	1.1	1.1	1.1
Middle East	**18,643**	**18,800**	**18,643**	**18,481**	**162**	**0.9**	**330.8**	**335.6**	**330.8**
Australia	577	545	577	521	56	10.7	82.0	82.0	82.0
Brunei	160	160	160	162	−2	−1.2	27.2	27.2	27.2
China	2,947	2,982	2,947	2,930	17	0.6	49.1	56.2	49.1
India	700	695	700	571	129	22.6	50.6	50.6	50.6
Indonesia*	1,340	1,340	1,340	1,330	10	0.8	161.1	161.1	161.1
Japan	15	16	15	16	−1	−6.3	8.0	7.7	8.0
Malaysia	680	630	680	650	30	4.6	61.2	61.2	61.2
New Zealand	45	45	45	42	3	7.1	14.3	14.3	14.3
Pakistan	56	56	56	58	−2	−3.4	42.3	42.3	42.3
Papua New Guinea	102	113	102	118	−16	−13.6	0.3	0.3	0.3
Thailand	56	53	56	51	5	9.8	31.8	31.8	31.8
Viet Nam	180	165	180	135	45	33.3	2.1	2.1	2.1
Other Asia-Pacific	23	22	23	23	—	—	25.8	25.7	25.8
Asia-Pacific	**6,881**	**6,822**	**6,881**	**6,607**	**274**	**4.1**	**555.8**	**562.5**	**555.8**
TOTAL WORLD	**60,908**	**61,575**	**60,908**	**60,492**	**416**	**0.7**	**7,087.9**	**6,984.9**	**7,087.9**
*OPEC	24,840	25,140	24,840	24,800	40	0.2	708.6	714.9	708.6
North Sea	5,432	5,690	5,432	5,041	391	7.8	413.8	400.6	413.8

*OPEC member. †Kuwait and Saudi Arabia production each include half of Neutral Zone. Source: Oil & Gas Journal. Data available in Oil & Gas Journal Energy Database.

RECENT OIL HISTORY

The world changed forever for the petroleum industry in the 1970s.

The Arab oil embargo of 1973-74, targeting mainly the U.S. for its support of Israel in the Arab-Israeli conflicts, resulted in the first of several oil price shocks that devastated the world economies and made petroleum the world's most widely watched commodity for the average consumer. In a matter of months, oil prices had increased more than fivefold to almost $12/bbl.

The Iranian revolution of 1978-79 squeezed oil supplies, again resulting in oil prices skyrocketing, more than doubling by 1980 to as much as $40/bbl.

After the string of price hikes had brought the real price of crude oil to a peak in 1981, oil prices started a trend of declines in 1982 that evolved into a full-fledged collapse in 1986. The average annual composite refiner acquisition cost of crude oil fell from $28.34/bbl in 1985 to $15.02/bbl in 1986. In many areas of the U.S. prices paid for crude oil at the wellhead dropped to as little as $7/bbl for lower quality crudes.

As was the case with the earlier price hikes that were implemented by OPEC in response to the supply disruptions, it was the move by key OPEC members, particularly Saudi Arabia, to recapture lost market share, that spurred the oil price collapse. The lost market share was the result of an unprecedented worldwide surge in oil production stemming from the oil price hikes of the 1970s.

Since 1986, oil prices have been marked by volatility. While they remained generally low during the late 1980s, it was yet another crisis in the Middle East that spawned a rollercoaster ride for crude oil prices. In 1990, the Iraqi invasion of Kuwait took more than 4.2 million b/d, or almost 7%, or world oil supply off the market, thus driving up the real, or inflation adjusted, price of crude oil to an average $19.61/bbl for the year. Prices for oil sold on the spot, or non-contract, market soared more than 50% to top $30/bbl at one point during the crusis. However, Saudi Arabia stepped up its production by more than 2 million b/d, along with increased output by other countries, the world drew on strategic stocks of oil supplies, and the shortfall was soon filled. Kuwait, devastated by Iraqi sabotage that almost obliterated its oil industry, has since rebuilt and is again producing more than 2 million b/d. Iraq remains at this writing under a United Nations embargo of its oil exports.

As oil producing nations demonstrated their ability to more than replace Iraqi and Kuwaiti oil supplies, coupled with a global economic recession that restrained oil demand, the real price of oil fell to $16.19/bbl in 1991. The continuing slackness in demand and surging production combined to drive real oil prices down to $15.22/bbl in 1992 and $13.21/bbl in 1993—the latter figure being the lowest annual average in 20 years. Economic recovery that gathered steam in 1994 and held steady in early 1995 bolstered demand for oil, but

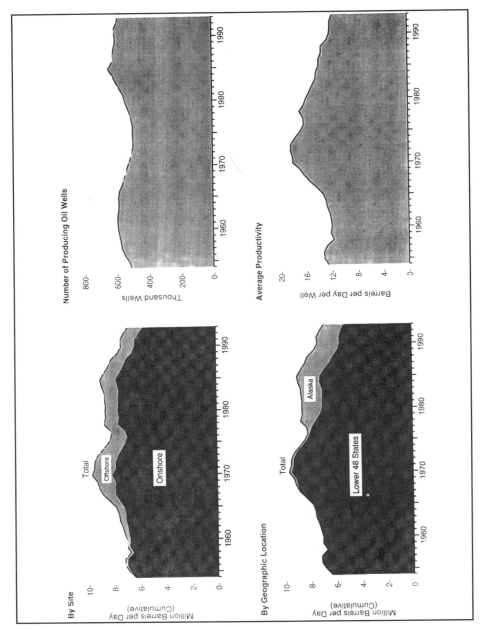

Fig. 1-7
U.S. crude oil and lease condensate production and oil well productivity.
Source: EIA.

continuing strength in production outside of OPEC—especially the North Sea—kept oil prices within a band of about $13-17/bbl for 1994.

Oil prices were stronger than many expected in early 1995, as the continuing surge in demand coupled with Iraqi oil sales still banned and OPEC production restraint kept crude at about $17-21/bbl for much of the first half. However, growing North Sea and other non-OPEC output together with the persistent specter of Iraqi oil returning to the market promised that 1995 would continue the pattern of oil price volatility.

RECENT NATURAL GAS HISTORY

As natural gas came to be regarded as a useful, clean source of energy instead of merely a waste byproduct of crude oil, its value grew. However, the relative difficulties in transporting a gaseous commodity when compared with liquids and an energy end-use sector infrastructure built largely around oil and coal kept natural gas from being a premium commodity for much of its early history. During these early years natural gas was an especially marginal commodity outside the U.S.

Throughout the 1950s and 1960s, the U.S. market for natural gas expanded as low prices encouraged demand. Of the many factors affecting natural gas markets during those decades, federal and state regulatory commissions were the most influential, keeping a tight ceiling on natural gas prices through a tiered structure for gas price mechanisms that was as confusing as it was inefficient in allowing producers to replace volumes produced by developing new reserves. Below-market rates for certain categories of natural gas, coupled with strong demand, ultimately resulted in regional shortages during the second half the 1970s.

In 1972, total U.S. consumption of natural gas reached a record of 22 trillion cubic feet (tcf). Thereafter, uncertainties about supply and rising energy prices began to erode demand. By the 1980s, lower demand resulted in a short term surplus of deliverable gas and production curtailments in many producing areas. In 1986, natural gas consumption totaled 16 tcf, the lowest annual total since 1965.

Weakened demand spanned all end-use sectors but was most severe in the industrial and electric utility sectors, where, particularly during the early and mid-1980s, the option of fuel switching proved to be most viable. Lower consumption by the industrial sector accounted for more than half of the 5.9 tcf drop during 1972-86, and lower use by electric utilities accounted for about one-fourth. After the low point in 1986, U.S. natural gas consumption has risen steadily, reaching more than 21 tcf in 1994, with all sectors increasing their use.

Natural gas supply and demand in the rest of the world also has jumped sharply during the past few decades. Vast supplies of natural gas in the offshore

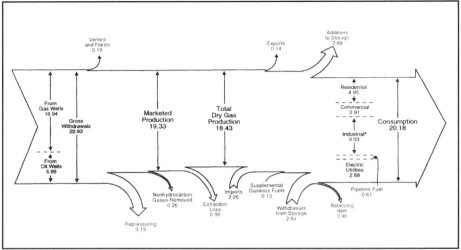

Fig. 1-8
U.S. natural gas flow, 1993. Source: EIA.

areas of Northwest Europe, especially the North Sea, in northern Africa, and in former republics of the Soviet Union supply growing demand in Europe. It was the development of technology to liquefy natural gas to contain huge volumes in specially built container ships for transportation across long distances overseas that prompted the creation of a huge market in Asia for natural gas. Far Eastern nations, notably Japan, thus are able to consume massive volumes of natural gas as regasified LNG that is sourced from Australia, Indonesia, Malaysia, Brunei, and the Middle East. Because of its clean burning qualities in a world increasingly concerned about environmental degradation, natural gas is expected to show the fastest growth of any major fuel in the next century.

Just as the oil price shocks of the 1970s changed the oil industry forever, the regulatory upheavals of the 1980s in the U.S. have forever changed the natural gas industry. During the 1980s, regulatory and legislative changes allowed consumers to purchase natural gas directly from producers and to arrange for pipeline and distribution companies to deliver it to them for a fee. Federal Energy Regulatory Commission (FERC) Order 636, implemented in November 1993, extended that trend toward a more efficient market by requiring interstate pipelines to unbundle, or separate, their sales and transportation services. In 1992, natural gas delivered for the account of others to industrial, electric utility, and commercial customers reached 7.4 tcf in the U.S., including 70% of deliveries to industrial customers, 64% to electric utilities, and 17% to commercial customers. Before long, U.S.

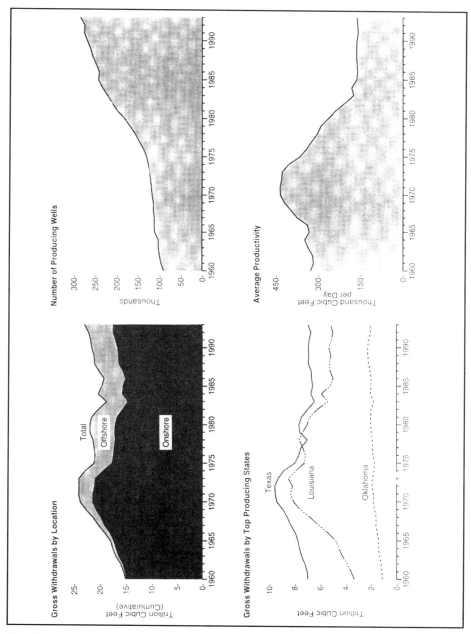

Fig. 1-9
U.S. natural gas gross withdrawals and gas well productivity. Source: EIA.

17

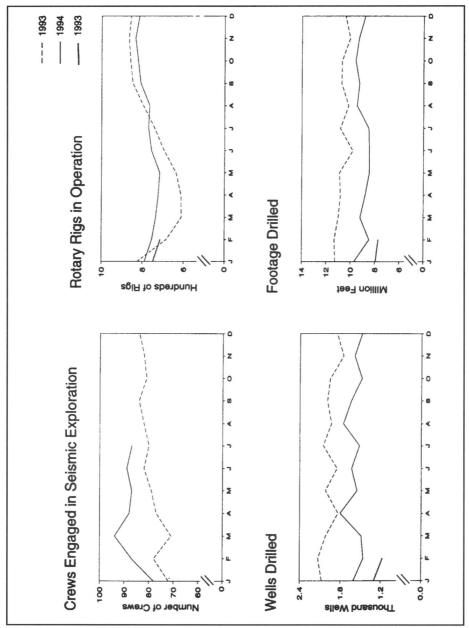

Fig. 1-10
U.S. oil and gas resource development indicators.

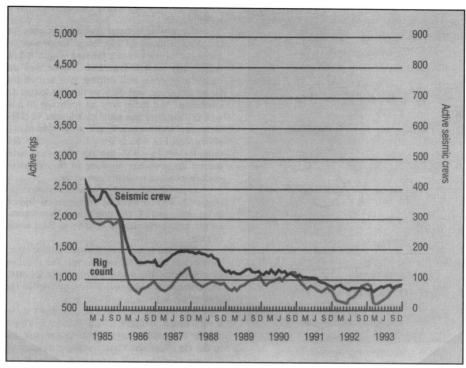

Fig. 1-11
U.S. active rotary drilling rigs, seismic crews.

pipelines' transportation vs. merchant sales of natural gas will account for all but a small fraction of such deliveries.

Another recent, significant market trend in the natural gas industry—again pioneered in the U.S.—is the increasing use of natural gas stored in depleted reservoirs or other subterranean features to help meet periods of sharply higher demand. The growing use of gas storage has helped to some degree flatten out the seasonal fluctuations in price that typically marked the industry in the past, when a severe cold snap might send the spot price of natural gas up by 100% or more. That in turn has helped producers' planning by providing a more predictable consumption rate but leaves production as less of a determinant in gauging the supply/demand balance.

In 1993, gross withdrawals of natural gas totaled 23 tcf, up for the seventh year in a row. That reflects an increasing role of natural gas produced from offshore wells and includes a small but rapidly growing amount of gas produced from coal seams, also known as coalbed methane. About three-fourths of all U.S. natural gas produced in the U.S. comes from gas wells, called dry or

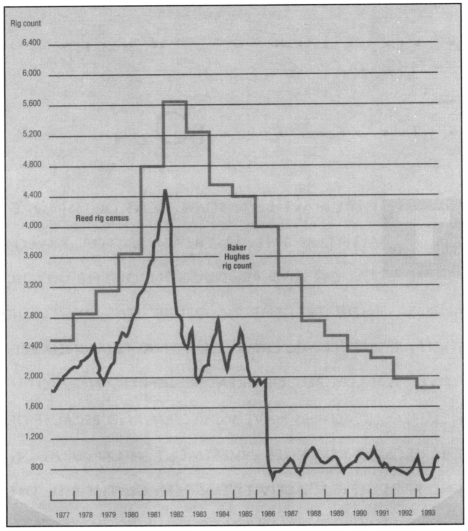

Fig. 1-12
U.S. rig fleet, utilization.

nonassociated gas, with the rest coming from oil wells, when it is referred to as associated gas.

As North America gas producers have found to their dismay, the creation of an essentially unfettered market in natural gas trade as a result of regulatory reform also meant the kind of price volatility oil markets have seen as well

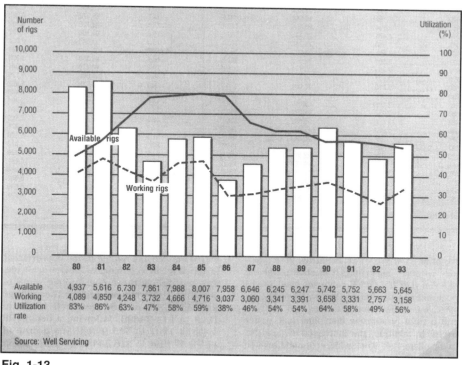

Fig. 1-13
U.S. well service rig utilization.

as gas-on-gas competition. The upshot is that gas prices have swung sharply in recent years, and analysts reckon the growing use of gas storage should help those swings flatten out.

Oil and gas producers alike watch a number of key market indicators that reflect the health of their industry, as Figs. 1-10 through 1-13 illustrate. These indicators show that boom times, while welcome by producers at the time, are unhealthy for the industry in the long run. When the perception of demand for any commodity or service exceeds for a lengthy time the perception of the supply for that commodity or service, costs spiral out of control while the consumers of those commodities and services rein their use and/or turn to alternatives. Thus, a crunch is inevitable between spiraling costs and declining demand. Inevitably, boom turns to bust. The global petroleum industry has at last recognized this, and the industry of the future is perpetually rethinking its mission and structure to ensure profitability amid the often wild vagaries of an uncertain marketplace, as we will see in the next chapter.

CHAPTER 2

Structure of the petroleum industry

We've learned a little bit about how the growing popularity of petroleum and its products has affected the world's economy during this century. Just as the supply and demand and markets for oil and gas have changed dramatically, so has the structure of the industry itself. Now let's take a look at the all important companies that find oil and gas, extract it from the ground, convert it into products we can use, and then market it (Fig. 2-1).

The primary distinction in the petroleum industry is between those companies that are integrated and those that are non-integrated. Integration refers essentially to a petroleum company that refines crude oil and markets the products (downstream) as well as produces crude and gas (upstream). The biggest of these companies are typically considered to be the "majors."

MAJORS

Most majors are headquartered in the U.S., but in recent years, the industry has seen the emergence and growth of major oil companies, many still partly or wholly owned by governments, in other countries.

Having upstream and downstream operations qualifies a company to be integrated, but only a handful of companies have the critical mass in terms of assets and breadth of markets to be considered majors. These, such as Exxon, Mobil, Chevron, Texaco, Amoco, and ARCO, have become household names and are among the biggest companies in the world.

According to the 1994 Oil & Gas Journal survey of the top 300 U.S. oil and gas companies, in terms of assets, these six companies hold more than half the total assets of the OGJ 300. If the list is extended to include the top 20 biggest companies in terms of assets, then those companies would have about 85% of the combined assets of the OGJ 300. The top 20 also tend to dominate in terms of oil and gas reserves and production.

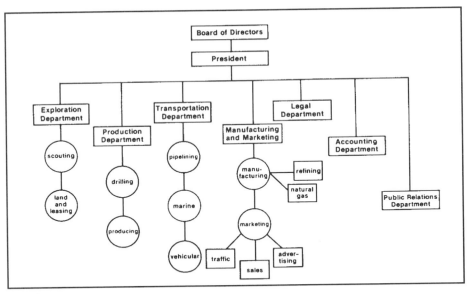

Fig. 2-1
Possible organization for an oil company.

There are majors headquartered outside the U.S. as well, including the biggest non-state owned petroleum company in the world, Royal/Dutch Shell. On a par with the U.S. majors in terms of critical mass and scope among the companies that are owned by public stockholders to a significant degree are British Petroleum, France's Total and Elf Aquitaine, Belgium's Petrofina, Australia's BHP, Canada's Imperial Oil, and Argentina's YPF.

Because natural gas is marketed primarily in a form not too dissimilar from how it is produced at the wellhead, having natural gas processing operations—which entails recovering liquids entrained in the natural gas stream—and even marketing these natural gas liquids (NGLs) generally has not been considered "downstream."

However, a new form of integration is emerging among natural gas companies in that some have natural gas production and marketing, processing, NGL marketing, pipeline transportation, local distribution, and end use sectors (usually gas-fired power plants)

What the industry is seeing now is the emergence of natural gas companies that consider themselves to be the gas equivalent of a major. These, such as British Gas, Enron, and Natural Gas, have the kind of critical mass and global reach that, while not in a league with an Exxon or Shell, certainly qualifies them to be viewed as majors in an industry with a rapidly growing albeit smaller worldwide scope.

Table 2-1

The Top 20 in Net Income and Stockholders' Equity, 1993

Rank	Company	Net income ($1,000)	Rank	Company	Stockholder's equity ($1,000)
1	Exxon Corp.	5,280,000	1	Exxon Corp.	34,792,000
2	Mobil Corp.	2,084,000	2	Mobil Corp.	17,237,000
3	Amoco Corp.	1,820,000	3	Shell Oil Co.	14,624,000
4	BP (USA)	1,461,000	4	Chevron Corp.	13,997,000
5	Chevron Corp.	1,265,000	5	Amoco Corp.	13,665,000
6	Texaco Inc.	1,068,000	6	Texaco Inc.	10,279,000
7	Conoco Inc.	812,000	7	ARCO	6,127,000
8	Shell Oil Co.	781,000	8	Occidental Petroleum Corp.	3,958,000
9	Enron Corp.	332,522	9	Meridian Oil Inc.	3,435,691
10	Union Pacific Resources Co.	309,000	10	Unocal Corp.	3,129,000
11	Meridian Oil Inc.	303,138	11	USX-Marathon Group	3,110,000
12	Occidental Petroleum Corp.	283,000	12	Amerada Hess Corp.	3,028,911
13	ARCO	269,000	13	Phillips Petroleum Co.	2,688,000
14	Sonat Inc.	261,240	14	Enron Corp.	2,623,374
15	Phillips Petroleum Co.	243,000	15	Coastal Corp.	2,278,100
16	Unocal Corp.	213,000	16	Consolidated Natural Gas Co.	2,176,432
17	Consolidated Natural Gas Co.	205,916	17	Kerr-McGee Corp.	1,512,000
18	Columbia Gas System	152,200	18	Pennzoil Co.	1,505,804
19	Ashland Oil Inc.	142,234	19	Ashland Oil Inc.	1,454,794
20	Pennzoil Co.	141,856	20	Sonat Inc.	1,363,215
	Total	17,427,106		Total	142,984,321

Source: Oil & Gas Journal, OGJ300

Table 2-2
The Top 20 in Spending and U.S. Net Wells Drilled, 1993

Rank	Company	Capital, exploratory spending ($1,000)	Rank	Company	U.S. net wells drilled
1	Exxon Corp.	8,167,000	1	Gerrity Oil & Gas Corp.	468.1
2	Chevron Corp.	4,440,000	2	Texaco Inc.	408.0
3	Mobil Corp.	3,656,000	3	Enron Corp.	379.5
4	Amoco Corp.	3,346,000	4	Mobil Corp.	365.0
5	Texaco Inc.	2,892,000	5	Chevron Corp.	350.0
6	Shell Oil Co.	2,279,000	6	ARCO	327.0
7	ARCO	2,070,000	7	Snyder Oil Corp.	326.8
8	Conoco Inc.	1,659,000	8	Union Pacific Resources Co.	323.0
9	Amerada Hess Corp.	1,348,041	9	Amoco Corp.	283.0
10	BP (USA)	1,250,000	10	Occidental Petroleum Corp.	282.6
11	Unocal Corp.	1,249,000	11	Meridian Oil Inc.	271.5
12	Phillips Petroleum Co.	1,216,000	12	Parker & Parsley Petroleum Co.	238.5
13	Occidental Petroleum Corp.	1,083,000	13	Exxon Corp.	231.0
14	USX-Marathon Group	910,000	14	Shell Oil Co.	226.0
15	Enron Corp.	695,437	15	HS Resources Inc.	218.9
16	Murphy Oil Corp.	637,556	16	Santa Fe Energy Resources Inc.	198.6
17	Freeport-McMoRan Inc.	568,197	17	Unocal Corp.	191.0
18	Meridian Oil Inc.	553,253	18	Sonat Inc.	175.1
19	Apache Corp.	543,500	19	Equitable Resources Inc.	152.7
20	Sonat Inc.	516,466	20	Cabot Oil & Gas Corp.	149.8
	Total	**39,079,450**		**Total**	**5,566.0**

Source: Oil & Gas Journal, OGJ300

INDEPENDENTS

Generally, non-integrated petroleum companies are considered "independents," whether their primary emphasis is upstream or downstream.

That includes a number of companies, such as Murphy Oil, Louisiana Land & Exploration, Tesoro, and Mapco, that have long been integrated to some degree but don't measure their assets in tens of billions of dollars.

Although independents are generally thought of as being primarily upstream companies, there are a number of sizable independent refiners and refiner/marketers, such as Citgo, Diamond Shamrock, and Tosco. The ranks of small independent refiners have thinned to a handful, however, as economies of scale in refining tend to accord survival to the larger refiners.

Economies of scale also had much to do with the economic devastation that hit the ranks of oil and gas producers during the 1980s when oil prices collapsed. While the larger independents and majors were hit by heavy losses, layoffs, restructurings, and consolidation, hundreds of small independent producers disappeared, usually in the rubble of bankruptcy.

Employment in the U.S. oil and gas extraction industry in the past decade has been cut in half, and the shrinkage continues today. According to the Independent Petroleum Association of America (IPAA), the annual average number of oil and gas employees in 1993 was 343,400, a decline of 10,400 from the 1992 average.

As the second edition of this book goes to press, a number of majors and independents are undertaking yet another round of restructuring and cutting, with employment upstream and downstream expected to decline again in 1995. IPAA figures that a sustained round of depressed oil and/or gas prices could put another 75,000 jobs at risk.

Typically, an independent producer is a person or corporation that produces oil and gas for the market but usually does not have a pipeline or refinery to move or process that product. Sometimes these operators are true entrepeneurs, leasing and drilling on small parcels of land that the majors either overlooked or didn't think worth fooling with—until a discovery is made.

Historically, independent producers have concentrated their efforts in the exploration and development areas. In fact, they are responsible for most exploratory wells. During 1969-78, independent producers drilled almost 90% of the new field widlcat wells in the U.S., while the majors accounted for the rest. The results were along the same lines: independents during the same period had more than 80% of the significant oil and gas discoveries, and the majors less than 20%.

Independents do most of their drilling onshore in the Lower 48 states, although there is a growing number of independents that are expanding their presence offshore. In the U.S., the majors generally drill the largely untapped

Table 2-3
The Top 20 in Liquids Reserves, 1993

Rank	Company	U.S. liquids reserves, million bbl	Rank	Company	Worldwide liquids reserves, million bbl
1	BP (USA)	2,561.0	1	Exxon Corp.	5,758.0
2	Exxon Corp.	2,324.0	2	Mobil Corp.	2,809.0
3	ARCO	2,259.0	3	ARCO	2,465.0
4	Shell Oil Co.	2,203.0	4	Chevron Corp.	2,414.0
5	Texaco Inc.	1,458.0	5	Shell Oil Co.	2,331.0
6	Chevron Corp.	1,279.0	6	Texaco Inc.	2,224.0
7	Amoco Corp.	1,256.0	7	Amoco Corp.	2,223.0
8	Mobil Corp.	1,116.0	8	Phillips Petroleum Co.	1,037.0
9	USX-Marathon Group	573.0	9	Conoco Inc.	964.0
10	Unocal Corp.	483.0	10	USX-Marathon Group	842.0
11	Phillips Petroleum Co.	432.0	11	Occidental Petroleum Corp.	793.0
12	Conoco Inc.	344.0	12	Unocal Corp.	764.0
13	Oryx Energy Co.	259.0	13	Amerada Hess Corp.	670.0
14	Santa Fe Energy Resources Inc.	230.9	14	Oryx Energy Co.	508.0
15	Pennzoil Co.	199.0	15	Maxus Energy Corp.	311.3
16	Amerada Hess Corp.	198.0	16	Santa Fe Energy Resources Inc.	248.2
17	Occidental Petroleum Corp.	195.0	17	Murphy Oil Corp.	202.4
18	Meridian Oil Inc.	168.2	18	Pennzoil Co.	201.0
19	Union Pacific Resources Co.	156.9	19	Kerr-McGee Corp.	199.0
20	Mitchell Energy & Development Corp.	122.7	20	Meridian Oil Inc.	168.2
	Total	17,817.7		Total	27,132.1

Source: Oil & Gas Journal, OGJ300

Table 2-4
The Top 20 in Liquids Production, 1993

Rank	Company	U.S. liquids production, million bbl	Rank	Company	Worldwide liquids production, million bbl
1	BP (USA)	228.9	1	Exxon Corp.	568.0
2	ARCO	221.0	2	Chevron Corp.	295.0
3	Exxon Corp.	202.0	3	Mobil Corp.	285.0
4	Texaco Inc.	155.0	4	ARCO	250.0
5	Shell Oil Co.	147.0	5	Amoco Corp.	236.0
6	Chevron Corp.	144.0	6	Texaco Inc.	228.0
7	Mobil Corp.	111.0	7	Shell Oil Co.	170.0
8	Amoco Corp.	100.0	8	Conoco Inc.	135.0
9	Unocal Corp.	48.0	9	Phillips Petroleum Co.	89.0
10	Phillips Petroleum Co.	47.0	10	Unocal Corp.	84.0
11	USX-Marathon Group	41.0	11	Amerada Hess Corp.	79.0
12	Conoco Inc.	40.0	12	Occidental Petroleum Corp.	79.0
13	Union Pacific Resources Co.	31.9	13	USX-Marathon Group	57.0
14	Amerada Hess Corp.	26.0	14	Oryx Energy Co.	43.0
15	Oryx Energy Co.	24.0	15	Union Pacific Resources Co.	31.9
16	Pennzoil Co.	24.0	16	Maxus Energy Corp.	27.9
17	Santa Fe Energy Resources Inc.	21.9	17	Santa Fe Energy Resources Inc.	24.3
18	Occidental Petroleum Corp.	21.0	18	Pennzoil Co.	24.0
19	Mitchell Energy & Development Corp.	20.3	19	Mitchell Energy & Development Corp.	20.3
20	Meridian Oil Inc.	15.3	20	Kerr-McGee Corp.	19.4
	Total	1,669.3		Total	2,745.8

Source: Oil & Gas Journal, OGJ300

Table 2-5
The Top 20 in Gas Reserves, 1993

Rank	Company	U.S. gas reserves, bcf		Rank	Company	Worldwide gas reserves, bcf
1	Amoco Corp.	11,767.0		1	Exxon Corp.	25,816.0
2	Exxon Corp.	9,530.00		2	Amoco Corp.	17,650.0
3	Chevron Corp.	5,484.0		3	Union Pacific Co.	17,321.2
4	Mobil Corp.	5,372.0		4	Mobil Corp.	16,959.0
5	Meridian Oil Inc.	5,221.0		5	ARCO	8,005.0
6	Shell Oil Co.	4,911.0		6	Chevron Corp.	7,741.0
7	ARCO	4,725.0		7	Unocal Corp.	6,632.0
8	Texaco Inc.	4,329.0		8	Phillips Petroleum Co.	6,069.0
9	Phillips Petroleum Co.	4,276.0		9	Texaco Inc.	5,970.0
10	Unocal Corp.	3,727.0		10	Meridian Oil Inc.	5,221.0
11	BP (USA)	2,806.0		11	Shell Oil Co.	5,199.0
12	USX-Marathon Group	2,044.0		12	USX-Marathon Group	3,748.0
13	Occidental Petroleum Corp.	1,980.0		13	Conoco Inc.	3,680.0
14	Anadarko Petroleum Corp.	1,836.0		14	Amerada Hess Corp.	2,653.0
15	Conoco Inc.	1,802.0		15	Occidental Petroleum Corp.	2,136.0
16	Union Pacific Resources Co.	1,731.2		16	Oryx Energy Co.	1,881.0
17	Pennzoil Co.	1,453.0		17	Anadarko Petroleum Corp.	1,875.0
18	Oryx Energy Co.	1,431.0		18	Enron Corp.	1,772.2
19	Enron Corp.	1,400.7		19	Pennzoil Co.	1,491.0
20	Mesa Inc.	1,202.4		20	Mesa Inc.	1,202.4
	Total	**77,028.3**			**Total**	**143,021.8**

Source: Oil & Gas Journal, OGJ300

Table 2-6
The Top 20 in Gas Production, 1993

Rank	Company	U.S. gas production, bcf	Rank	Company	Worldwide gas production, bcf
1	Amoco Corp.	867.0	1	Mobil Corp.	1,665.0
2	Chevron Corp.	751.0	2	Exxon Corp.	1,583.0
3	Exxon Corp.	697.0	3	Amoco Corp.	1,487.0
4	Texaco Inc.	652.0	4	Chevron Corp.	902.0
5	Mobil Corp.	558.0	5	Texaco Inc.	748.0
6	Shell Oil Co.	539.0	6	Unocal Corp.	623.0
7	Unocal Corp.	365.0	7	Shell Oil Co.	553.0
8	Phillips Petroleum Co.	345.0	8	Phillips Petroleum Co.	509.0
9	Meridian Oil Inc.	336.0	9	Conoco Inc.	481.0
10	ARCO	332.0	10	ARCO	449.0
11	Conoco Inc.	305.0	11	Meridian Oil Inc.	336.0
12	Enron Corp.	240.0	12	Amerada Hess Corp.	323.0
13	Union Pacific Resources Co.	226.0	13	USX-Marathon Group	317.0
14	Pennzoil Co.	220.0	14	Enron Corp.	262.2
15	Occidental Petroleum Corp.	219.0	15	Occidental Petroleum Corp.	238.0
16	USX-Marathon Group	193.0	16	Union Pacific Resources Co.	226.0
17	Oryx Energy Co.	191.0	17	Pennzoil Co.	223.0
18	Amerada Hess Corp.	183.0	18	Oryx Energy Co.	220.0
19	Anadarko Petroleum Corp.	159.0	19	Anadarko Petroleum Corp.	162.0
20	Sonat Inc.	146.1	20	Sonat Inc.	146.1
	Total	**6,657.1**		**Total**	**11,453.2**

Source: Oil & Gas Journal, OGJ300

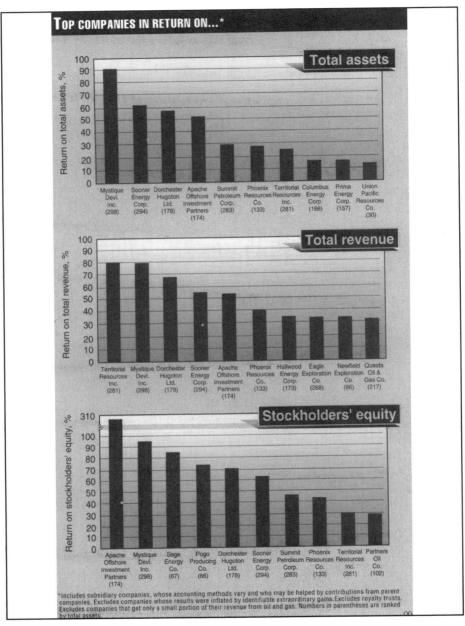

Fig. 2-2
Independents dominate the list of the top performing oil and gas companies in return on investment. *Source: Oil & Gas Journal, OGJ300*

regions offshore and in Alaska, where the focus is on discoveries with much greater potential reserves. Consequently, the majors' 20% of discoveries during the 1969-78 study period accounted for 44% of the oil and gas reserves found and while the independents' much more numerous discoveries accounted for only 56% of the oil and gas reserves found.

In absolute terms, independents' oil production has remained fairly steady since the late 1970s. As a percentage of total U.S. production, the independents' share rose in the early 1980s but has remained little changed since then. Declines in U.S. production were borne for the most part by the major oil companies. In fact, if we ignore oil production from Alaska's North Slope, the only factor preventing U.S. crude oil production from declining precipitously has been the activity of independent producers. The importance of the independent producers' role in preventing a catastrophic fall in U.S. oil and gas production is further illustrated by their share of oil and gas well completions, which increased during the boom years of 1974-82 from 68.5% to 85%.

These ratios essentially hold true today. In its 1995 forecast, Oil & Gas Journal predicts that majors will drill about only 3,200 of the roughly 33,000 wells to be drilled in 1995 in the U.S. and western Canada. That is the same ratio seen in 1994.

This program of exploration and development has been sustained by the relationship between price and expenditures. Since 1971, the trend has been rising expenditures for exploration and development that roughly parallel increases in price, attributable to the world oil market and the intrastate natural gas market. Exceptions occur when there is a significant change in tax treatment of either expenditures (e.g., minimum tax on intangible drilling costs) or revenue (e.g., percentage depletion).

Producers generally have income from both crude oil and natural gas. Thus, their income (and thus funds available for reinvestment) depends on the composite price of oil and gas. Accordingly, expenditures to develop petroleum resources are much a function of the price of oil and gas. As prices rise, so do expenditures for exploration and development. If prices remain relatively constant, so too do expenditures for exploration and development.

Remember that the benefits of increased exploration and development are never immediately evident. There is a lead time—sometimes substantial—between expenditures and actual production. In highly controversial or high-cost operating areas such as Offshore California or Alaska, that lead time can sometimes be measured in decades.

The independent producer is important in developing the domestic resources of his country, whether the U.S. or another country. Independents are very sensitive to fluctuations in wellhead prices and to increases in operating costs that come from increased government regulation and taxes. Therefore, policies that affect the oil and gas industry must be evaluated carefully. Wherever operating costs in a region or country rise unexpectedly and independently of changes in wellhead prices, majors, with their global breadth of operations and economies of scale, can choose to reduce their exposure in the affected area and focus on friendlier investment climates or another aspect of their operations, such as downstream. Most independent producers don't have the luxury of those options. That is why the petroleum industry is becoming an increasingly global one.

A GLOBAL INDUSTRY

After the rapid expansion and then contraction of the petroleum industry during the 1970s and 1980s, the decade of the 1990s has been marked by two key new trends: globalization and privatization, with the latter fueling the former.

For a variety of reasons, primarily the lack in the U.S. of potential new giant and "elephant" class discoveries that majors favor, the high U.S. operating costs, and the increasingly onerous regulatory burden in the U.S. (especially in the environmental arena), there has been a massive exodus of upstream capital from the U.S. to other countries. Advances in technology, especially computer-based, have uncovered new geological theories that make areas previously dismissed or overlooked attractive for exploration again. At the same time, many more countries are sweetening incentives for exploration and development to attract foreign capital and expertise than has ever been the case in the past.

These two trends have resulted in literally breathtaking turnarounds in the economic welfare of some countries. Several examples come to mind. Colombia, which not too long ago was a net importer of oil, is evolving into one of the Western Hemisphere's biggest oil exporters as a result of major oil discoveries in mature producing regions by foreign oil companies. Desperately impoverished nations with essentially no commercial hydrocarbon production, such as Papua New Guinea and Vietnam, almost overnight have become significant oil exporters and are rapidly building a petroleum infrastructure on the strength of major oil and gas discoveries in recent years by foreign majors. These kinds of examples have spurred other countries to follow suit with improved terms for foreign investment, and there now is a scramble on as nonproducers, new producers, and mature producing nations recognize they must compete for that multinational petroleum capital and expertise. It isn't just the majors in this hunt, either. Increasingly, many independents are recognizing that a global approach is essential not just to prosperity but also to survival in the petroleum industry of tomorrow. With their greater speed of movement and flexibility, small independents these days often are authoring geological "plays" in certain areas of the world, only to bring in majors and large independents later to help with capital and clout.

Table 2-7

OGJ100: Reserve and Production Leaders, 1993

Rank	Company	Production, million bbl	Rank	Company	Reserves, million bbl
1	Saudi Arabian Oil Co.	2,950.0	1	Saudi Arabian Oil Co.	258,703.0
2	National Iranian Oil Co.	1,327.5	2	Iraq National Oil Co.	100,000.0
3	China National Petroleum Co.	1,061.4	3	Kuwait Petroleum Corp.	94,000.0
4	Petroleos Mexicanos	975.7	4	National Iranian Oil Co.	92,860.0
5	Petroleos de Venezuela SA	894.3	5	Abu Dhabi National Oil Co.	92,200.0
6	Royal Dutch/Shell	747.0	6	Petroleos de Venezuela SA	64,447.0
7	Nigerian National Petroleum Corp.	692.1	7	Petroleos Mexicanos	50,925.0
8	Kuwait Petroleum Corp.	682.2	8	China National Petroleum Co.	24,000.0
9	Abu Dhabi National Oil Co.	654.1	9	National Oil Corp. (Libya)	22,800.0
10	National Oil Corp. (Libya)	499.7	10	Nigerian National Petroleum Corp.	17,899.8
11	Pertamina (Indonesia)	484.0	11	Sonatrach (Algeria)	9,200.0
12	British Petroleum Co.pic	453.3	12	Royal Dutch/Shell	8,744.0
13	Ministry of Petroleum and Minerals (Oman)	284.4	13	Oil & Natural Gas Commission (India)	5,920.9
14	Sonatrach (Algeria)	274.9	14	Pertamina (Indonesia)	5,779.0
15	Petronas (Malaysia)	235.4	15	Ministry of Petroleum and Minerals (Oman)	4,700.0
16	Petroleo Brasileiro SA (Brazil)	232.9	16	British Petroleum Co.pic	4,537.0
17	Elf Aquitaine	226.0	17	Petronas (Malaysia)	4,300.0
18	Syrian Petroleum Co.	210.3	18	Dubai Petroleum Co.	4,000.0
19	AGIP	195.0	19	Qatar General Petroleum Corp.	3,729.0
20	Oil & Natural Gas Commission (India)	194.9	20	Petroleo Brasileiro SA (Brazil)	3,600.0
	Total	872,344.7		Total	13,275.1

Source: Oil & Gas Journal

At the same time, there has been a growing trend in the world toward democratization and privatization, most dramatically reflected in the collapse of Communism in the former Soviet Union (FSU) and eastern Europe. There is now almost a universal acknowledgment that the countries that will become stable and prosperous are those that are at least moving towards democratic and free market principles.

That acknowledgment underlies the burgeoning privatization of petroleum sectors in a number of countries. For the first time in decades for some countries, including some of the key OPEC producers and even the remaining Communist states, foreign petroleum companies are being welcomed to not only offer their services but also to take equity stakes in upstream, midstream, and downstream operations. Even countries that nationalized their petroleum sectors in recent decades, such as Venezuela, are reversing that process. In many instances, even ages-old monopolies in the petroleum sector, such as Italy's, are being dismantled as companies and assets are sold piecemeal or even whole in recognition that government involvement in the oil gas sector ensures inefficiencies. It's a sensitive area for many governments, which must often dismantle domestic subsidies for hydrocarbon based fuels, with resulting price spikes causing civil discord, as in much of Latin America, or must sell what generations have considered to be the "national patrimony," as is the case in Mexico.

The globalization flows both ways. Part of this trend includes the growing significance of state-owned and formerly state-owned petroleum companies (Table 2-7). In some instances, such as YPF's, total privatization has created a new major overnight with the assets and now the flexibility and freedom to operate that makes it a force to be reckoned with in world markets. In other cases, such as Kuwait's and Saudi Arabia's state petroleum companies, there is a growing recognition that they must look beyond their own borders and emulate the global breadth and integration of the public stock companies while retaining those companies' flexibility.

As these trends continue, the shape of the world petroleum industry will continue to change dramatically. There will be many new players, as a host of new oil companies are springing up in the FSU in the place of former state monopolies. Russia alone is likely to produce a number of integrated companies with the assets and scope to rightfully be called majors.

Restructuring, cost containment, globalization, privatization. These are no longer buzzwords of the 1990s for the petroleum industry. They have become essential components of the corporate culture and structure that is the petroleum industry today.

Enough said on the structure of the petroleum producing industry. Let's jump into the fascinating world of petroleum production now and learn what kind of formation a producing company looks for in its search for oil and gas.

CHAPTER 3

Productive Formation

The earth's crust is composed essentially of three types of rocks: igneous, metamorphic, and sedimentary. Although oil and gas are found in all three kinds of rock, they are most closely associated with sedimentary rock. Sedimentary rocks come from a variety of sources but in general were laid down on the earth by the action of wind or water, or through chemical deposition (e.g., leaching). These sedimentary materials can be classified as (1) rocks, (sandstone, shale), (2) carbonates (certain limestones), and (3) dolomites.

Although sedimentary rocks are associated with oil, not all sedimentary rocks contain oil. In order for petroleum to be present, most scientists theorize that the remains of plant and animal life, as well as the presence of certain temperatures and pressures, were needed. So how did this environment occur?

Fig. 3-1
As plants and animals die, they fall to the seafloor and are buried in sand, silt, and mud.

Early life began in vast seas and inland lakes that covered large portions of the present continents. As the abundant populations of marine plant and animal life died, their remains were buried rapidly and preserved in the silt and mud that continuously filtered down to the ocean floor (Fig. 3-1).

Rivers carried great volumes of mud and sand to be spread by currents and tides over the ever-changing sea shoreline. This joined the marine life remains that settled to the bottom of the sea and deltas and were repeatedly buried. The mud and seawater protected the material from further decay. As more and more layers of organic material, sand, silt, clay, and lime accumulated and time passed, the weight of the overlying sediments exerted great pressure on the deeper sedimentary layers. With the increasing weight of the accumulating sediments, the seafloors slowly sank, forming and preserving thick sequences of mud, sand, and carbonates. These eventually formed into sedimentary rocks. The tremendous pressure—along with the high temperature, bacterial action, and chemical reactions—caused the formation of crude oil and natural gas.

ACCUMULATION AND OCCURRENCE

Contrary to popular belief, oil and gas do not exist in great lakes and rivers below the surface of the earth (even though we do talk about petroleum reservoirs). Instead, hydrocarbons—crude oil and natural gas derived from the restructured hydrogen and carbon in the remains of early plant and animal life—occur as fluids that occupy the pore spaces of the sedimentary rocks.

Going back to our sedimentary environment, the layers of silt that originally contained the decaying plant and animal material are known as *source beds*. These source beds include the dark marine shales and marine limestones. Continued squeezing of the source beds where the sediments were deposited and transformed resulted in pressures and temperatures high enough to let the oil and gas *migrate* out of the source rocks and accumulate in the adjoining porous and permeable reservoir rocks (Fig. 3-2), such as sandstones, carbonates (limestones), and dolomites. These latter rocks are repositories for the migrating hydrocarbons and are known as *reservoir rocks*.

But how can oil and gas move through rock? Isn't rock solid? Not really. Rock is filled with millions of tiny spaces and interstices called *pores* (Fig. 3-3). These pores are the spaces between the individual grains that make up the reservoir rock. Some rocks have large pores; others have small pores. The ratio of the pore volume to the total rock volume is known as porosity, commonly expressed as a percentage. A good sandstone may have as much as 30% porosity, while a tight limestone may have as little as 5% porosity. So the greater the percentage of pore volume, the greater the capacity a rock has to hold large quantities of petroleum.

Fig. 3-2
The weight of overlying rock layers compacts the seafloor, squeezing the hydrocarbons out of the source beds and upward into the reservoir rocks.mud.

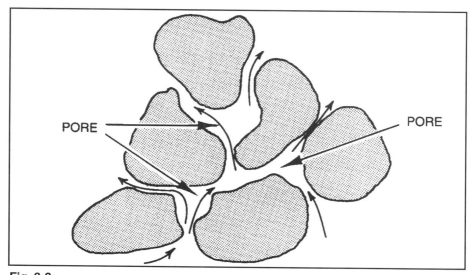

Fig. 3-3
Petroleum collects in spaces, or pores, in rocks and flow through the tiny cracks and channels. This is called porosity and permeability.

In addition to finding available pore space, the hydrocarbons must be able to move from pore to pore and eventually migrate up closer to the surface. The ease with which fluid moves through the interconnected pore spaces of rock is called permeability. The higher a rock's permeability, the easier it is for hydrocarbons to move from pore to pore within the rock.

To illustrate the concept of porosity, let's use the following example. Take two buckets of equal volume. Fill one with dry sand and the other with water. Then slowly empty the bucket of water into the bucket of sand. If we can empty the entire bucket of water into the sand without overflowing, we have 50% porosity. If we can empty only half the bucket, we have 25% porosity, etc.

So the tiny spaces or pores between particles in the sediments provide the openings in which oil and gas can accumulate. And the tiny cracks in the rocks let the oil and gas migrate from the source rocks and through the reservoir rocks.

Petroleum migration appears to occur in two separate stages. First, hydrocarbons are lighter than water. If you place a drop of motor oil in a pan of water, it floats on the surface of the water. Likewise, oil and gas moved up from the lower seafloor source beds where they were generated into the more porous rocks above. Within the porous layers, they continued to move up until they reached a layer of nonpermeable rock, which trapped the fluids.

What do we mean by layers and traps? Remember that sedimentary rocks are deposited in essentially horizontal layers or shallow slopes called *strata* or beds (See Fig. 3-1). As additional layers were deposited, the lower layers were squeezed and compacted to form rock. However, most rock layers are not strong enough to withstand movements and pressure of the earth's crust, so they deform.

One kind of deformation is *folds*, which is usually the force behind mountain chains like the Rockies (Fig. 3-4). Folds range in size from small wrinkles to great arches and troughs many miles across. The upfolds or arches are called *anticlines*; the downfolds or troughs are called *synclines*. Folds may be symmetrical, with similar flank dips on both sides, or asymmetrical, with one limb steeper than the other. A very short anticline whose crest plunges in opposite directions from a high point is called a *dome*. Domes are important to the oil industry because those were the first geological structures discovered to trap oil and gas.

Faults are another kind of deformation. Nearly all rocks are fractured to some extent and form cracks called *joints*. If the rock layers on one side of a fracture or joint move in a different direction from the layers on the other side of the joint, that constitutes a fault. These faults can displace rock layers from only a few inches to many thousands of feet and sometimes even miles, such as along the San Andreas fault in California. Faults are usually classified as either normal, reverse, thrust, or lateral, depending on the movement (Fig. 3-5).

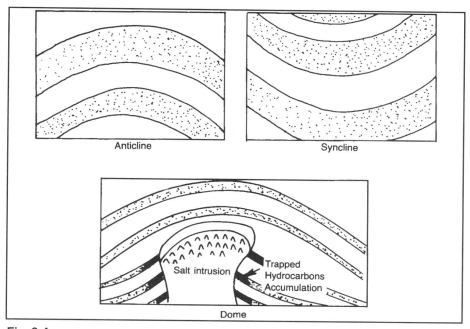

Fig. 3-4
Some typical kinds of folds.

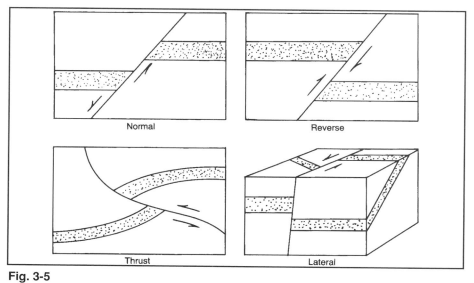

Fig. 3-5
Faults, another kind of deformation.

Fig. 3-6
Unconformities.

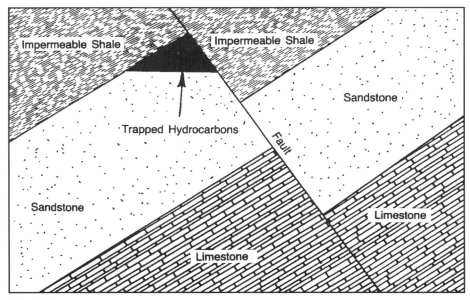

Fig. 3-7
A combination trap. The hydrocarbons are trapped structurally by the fault and stratigraphically by the impermeable shale layer.

Movement is upward or downward in normal and reverse faults but mainly horizontal in thrust and lateral faults. Faults may also have a combination of vertical and horizontal movements.

Another result of earth movement is to erase or to prevent the deposition of part of a series of sediments that are present elsewhere. This buried erosion surface is called an unconformity (Fig. 3-6). This is often an important structure because of its trapping capabilities.

Earth movements are very important to the study of petroleum geology because they produce barriers that contain a large proportion of petroleum accumulations. Remember that oil and gas continue to migrate, ever moving up—sometimes vertically, sometimes laterally—until they are finally trapped by some kind of deformation in the strata or layers of rock.

Traps are classified into three major types: structural, stratigraphic, and combination. *Structural traps* are where oil and/or gas is localized as the result of a structural condition (a fold or fault) in the reservoir rock. This condition is caused by the movement of the earth's crust. *Stratigraphic traps* are where oil and/or gas is localized as the result of variations in the rock's lithology, i.e., changes in rock type or porosity. And *combination traps* include features of both structural and stratigraphic traps (Fig. 3-7).

Therefore, three things are required for petroleum to accumulate. First, there must be a source of oil and gas. Second, there must be a reservoir rock, a porous bed permeable enough to let the fluids flow through it. Third, there must be a trap or barrier to stop fluid flow so accumulation can occur. The next stage in the process is segregation of the fluids in the reservoir.

OIL AND GAS SEGREGATION AND RESERVOIR DRIVES

When petroleum migrates into a trap, it displaces salt water left by the ancient seas. Petroleum floats on salt water as easily as it does on fresh water (big oil spills are evidence of this). So the oil and gas continue to migrate up, leaving the salt water in the lower section of the reservoir (Fig. 3-8). Gas is even lighter than oil, so it is usually found in the highest portion of the trap. Oil and oil with dissolved gas are found below the gas. Salt water is below the oil.

Although salt water is heavier than oil, not all of it is completely displaced from the pore spaces in the trap. This remaining water, called *connate water*, fills the smaller pore spaces and coats or forms a film over the surfaces of the rock particles or grains. The oil and/or gas occupies these water-coated pore spaces. That's why salt water is often produced along with oil and gas in a well. As the oil and gas flow to the wellbore and up to the surface, they carry along the connate water.

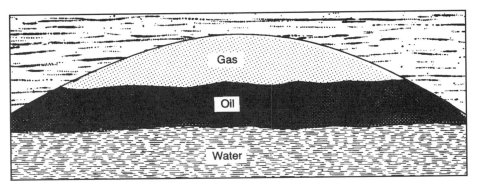

Fig. 3-8
Gas, oil, and water tend to divide themselves in reservoirs according to their densities.

What creates the driving force that moves the fluids in the rock to the wellbore? Sometimes it's a difference in pressure. Fluids move from areas of high pressure to areas of low pressure. The wellbore has less pressure than the crushing layers of surrounding rock, so the oil, gas, and water flow toward it.

Often water will contribute to this movement. If the pressure is released at the top of a reservoir, the water will force the overlying layers of oil and gas to push upward into the wellbore. This is called *water drive*. A *gas cap drive* works similarly. Gas is associated with oil and water in reservoirs in two principal ways: as *solution gas* and as *free gas*. Natural gas remains in solution if the pressure is sufficiently high and the temperature is sufficiently low. When the oil reaches the surface and the pressure is relieved in separating equipment, the gas comes out of solution. Free gas tends to accumulate in the highest structural part of the reservoir and forms a gas cap. In a gas cap drive, the wellbore is drilled into the oil layer. As the oil is depleted, the gas cap expands to relieve pressure and continues to push the oil into the wellbore. (See Chapter 4 for more about drives.)

Having gas in solution is advantageous when producing a well. As long as there is free gas in a reservoir gas cap, the oil in the reservoir will remain saturated with gas in solution. And having gas in solution lowers the viscosity (or flow capability) of the oil, making it easier to move to the wellbore.

CLASSES OF PETROLEUM

One of the main ways to classify petroleum is API gravity. API gravity is a name given to a measurement found in a formula set up by the American Petroleum Institute (API). The main factors that seem to affect the gravity of crude oil are the formation temperature and pressure. In most sedimentary basins, the oil becomes lighter and the API gravity becomes higher as depth increases. Older, deeper rocks usually have higher API gravity ratings, while younger, shallower rocks generally have lower API gravity ratings. These ratings are important in evaluating the saleability of a particular grade of petroleum.

Another important point in classifying petroleum for sale is the amount of impurities in the oil and gas. Impurities occur as free molecules or as atoms attached to the larger hydrocarbon molecules. The most common impurity found associated with crude oil and gas is sulfur. Sulfur is very corrosive and must be refined in special refineries. Therefore, petroleum that contains sulfur brings a lower price per than petroleum that is relatively free of sulfur. In addition, sulfur becomes a hazard in drilling. As hydrogen sulfide, it becomes a deadly gas that can kill a person in as little as 10 seconds.

The origin, migration, and accumulation of petroleum is a highly inefficient process. Only about 2% of the organic matter dispersed in fine-grained rocks becomes petroleum, and only about 0.5% will accumulate in a reservoir for commercial production.

There are approximately 200 times as many dispersed hydrocarbons worldwide as there are reservoirs of hydrocarbons. This is in part because the volume of reservoir rock is smaller than the total sediments in the earth's crust. Within prospective parts of oil-forming basins, the ratio varies between 10:1 and 100:1.

Finally, several conditions must be met before a reservoir becomes productive:

- A trap must exist to block the oil and gas.
- A reservoir must have enough thickness, areal extent, and pore space to accumulate an appreciable volume of hydrocarbons.
- It must be able to yield the contained fluids at a satisfactory rate after the reservoir is opened to production.
- Most important of all, it must contain enough fluids to make the entire venture commercially worthwhile.

Once all these conditions are met, the reservoir is ready to be developed.

CHAPTER 4

Development

We've learned what it takes to form a commercially viable reservoir of petroleum. Now let's take a look at the steps involved up on the surface in order for a company to drill a wellbore into the earth to test and produce a formation. It all begins in the land department.

LAND DEPARTMENT FUNCTIONS

In departmentalized operations, undeveloped properties—properties where no wells have been drilled—are managed in the land department. Working closely with the land department are the exploration and legal departments. In fact, the roles of the exploration and land departments are so closely intertwined that the two divisions are often combined.

The land department acquires undeveloped properties and services them until the land is produced or until a company decides to dispose of a property. The land department is also usually concerned with the extent to which a company is immobilized in unproductive properties. The exploration department recommends which properties should be acquired, retained, developed, or abandoned. And the legal department examines titles, conducts any necessary title litigation, and approves or drafts legal documents.

When acquiring undeveloped properties, the land department acts upon information from two different sources: scouts and information exchanges. The scout force keeps the company posted on competitors' exploration and leasing activities.

Information exchanges, such as swapping well logs with another company, also help reduce the secrecy that petroleum companies once used to clothe their operations. In fact, general information on exploration and leasing activities is now available from land survey information services.

The information obtained by the scouts may lead to the acquisition of "protective acreage" in areas where the company has not yet explored. Or the general information may give clues to geologists that an existing field may not yet be fully developed. Whatever the case, once a prospect is located, the land-man steps in.

A *landman* negotiates directly with landowners or lease brokers to acquire acreage. Like the scout, the landman needs a wide range of knowledge as well as the ability to deal effectively with people. He or she must have a general background of the petroleum industry as well as a working knowledge of contract law, property law, accounting, taxation, and state regulations.

During the leasing stage, the landman offers a lease to a landowner, who usually reserves a one-eighth royalty interest of the total oil and gas production (Fig. 4-1). Almost invariably, the oil company has to pay the landowner a bonus, too, to sign the lease. These bonuses vary from a promise to drill a well within a specified period of time to a cash payment on a per-acre basis. Whatever is agreed upon, the rights must be obtained from the landowner before a well can be drilled on a property.

Even before drilling commences, though, the landowner may be involved in negotiations. Frequently an oil company will want to send a geophysical crew to survey the prospect. Under this option, the company usually buys the right to perform the survey and to lease the land at a fixed price if a drillable prospect is uncovered. Generally the landowner asks for a fee per shot hole.

Often, the surface landowner owns the mineral rights beneath the surface of his property. Occasionally, though, two different owners are involved—one with surface rights and one with mineral rights. The process becomes even more complicated when tracts of land are divided and given or sold to several parties. So the landman's search can become exhaustive, hunting for the appropriate owners' approvals for mineral as well as surface rights.

What if a prospect is on land owned by the government—public domain? If this is the case in the U.S., the exclusive privilege of testing and leasing may be secured from the Bureau of Land Management (BLM) under the Mineral Land Leasing Law of 1920. The BLM, a subagency of the Department of Interior, manages operations on those lands. Another Interior agency, Minerals Management Service, provides a similar function offshore.

Whichever the case—private land or public domain—before any development work is begun, an operator must acquire title to the land to be tested or conclude a lease with a landowner that gives the operator rights to conduct exploratory testing and drilling operations on the land. The operator also acquires rights to produce, sell, and remove any oil, gas, or other minerals that may be discovered on the property. And the operator tries to acquire the rights on adjacent properties. More about that will follow in the sections on field development. First, before a field can be developed, a drilling company moves in to drill a test hole—the next phase in operations.

RECORDING REQUESTED BY

WHEN RECORDED MAIL TO

SPACE ABOVE THIS LINE FOR RECORDER'S USE

OIL, GAS AND MINERAL LEASE

THIS LEASE AND AGREEMENT, made and entered into_____, 19____

by and between_____

hereinafter called "Lessor" (whether one or more), and _____, hereinafter called "Lessee."

WITNESSETH: For and in consideration of a rental paid in advance upon execution hereof, receipt of which is hereby acknowledged, and the covenants and agreements hereinafter contained on the part of Lessee to be kept and performed, Lessor does hereby grant and lease to Lessee the land hereinafter described (herein sometimes called the "leased land") for the purposes and with the exclusive right of prospecting, exploring, mining, drilling and operating the leased land for oil, gas, other hydrocarbons, associated substances, sulphur, nitrogen, carbon dioxide, helium and other commercially valuable substances which may be produced through wells on the leased land, whether or not similar to the above-mentioned substances (hereinafter collectively called "substances") and producing, taking, treating, storing, removing and disposing of such substances from the leased land, together with the right to construct, erect, maintain, operate, use, repair, replace and remove pipelines, telephone, telegraph and power lines, tanks, machinery, appliances, buildings and other structures, useful, necessary or proper for carrying on its operations on the leased land, the right to drill thereon for water and the free use of water so obtained (but not water from Lessor's wells in operations on the leased land, and rights-of-way for passage over, upon and across and ingress and egress to and from the leased land for any or all of the above-mentioned purposes. Any pipelines, pole lines or roads so constructed by Lessee may also be used by it in its operations on lands in the vicinity of the leased land. Lessor shall have the right to occupy and use the leased land in any manner and to any extent not inconsistent with Lessee's rights or in interference with Lessee's operations hereunder. The land hereby leased is situated in the county of_____, state of California, and is described as follows:

together with such rights as Lessor may have in any roads, streets, alleys, waterways, canals, sloughs, levees, ditches, easements, rights and rights-of-way upon, within or adjoining the above-described property and containing_____acres, more or less.

TO HAVE AND TO HOLD the same for a term of 20 years from and after the date hereof and so long thereafter as Lessee shall conduct development (including, without limitation, drilling, redrilling, deepening, repairing and reworking) or producing operations on the leased land or lands pooled therewith without cessation for more than 90 consecutive days, or be excused therefrom as hereinafter provided.

In consideration of the premises, the parties hereby agree as follows:

1. On or before_____years after the date hereof (the last day of said period being hereinafter referred to as the "working date"), Lessee shall either commence drilling operations on the leased land and thereafter continue its operations with reasonable diligence until oil or gas or another of said substances is found in paying quantities or a depth is reached at which further drilling would, in the judgment of Lessee, be unprofitable, or quitclaim and surrender this lease as hereinafter provided.

2. Lessee has paid Lessor rental in full hereunder for the first_____months of the term hereof. If Lessee has not commenced drilling operations on the leased land or terminated this lease within that time, Lessee, commencing with the expiration thereof, shall pay or tender to Lessor annually in advance, as rental, the sum of_____per acre for so much of the above-described land as may then still be held under this lease at the time of payment and shall continue such payments until drilling operations are commenced or this lease terminated.

3. The payments required to be made by Lessee hereunder may be made by its check issued and made payable as hereinafter provided. All persons entitled to participate or share in such payments shall, at the request of Lessee, unite in a written designation of one person, bank or corporation as Lessor's agent to receive such payments, to the end that Lessee shall not be required to make any payment otherwise than by one check, which check shall be payable to but one payee, such payee to assume the burden and responsibility of making a proper distribution without expense to Lessee among the persons entitled thereto. When such designation is made, said payments may be made by mailing such check to the payee at the address designated. Until such designation is made, such checks may be made payable and may be mailed to_____

A waiver by Lessee of the provisions of this paragraph in the making of any payment or payments shall not be deemed a waiver thereof with respect to subsequent payments. If at any time there be no one person, bank or corporation authorized to receive payments hereunder, the time for making such payments shall be extended until Lessee has been notified of such designation.

4. Any notice to be given by either party to the other hereunder may be delivered in person or by registered or certified mail, postage prepaid, addressed to the party for whom intended as follows: to Lessor at_____

to Lessee at_____

_____Either party may from time to time, by written notice to the other, designate a different address which shall be substituted for the one above specified. If any notice from one party to the other is given by registered or certified mail, usual time for transmission of mail shall be computed and at the end of such time service of notice will be considered made.

Fig. 4-1
Sample oil, gas, and mineral lease.

DRILLING OPERATIONS

The geologists and geophysicists have examined maps and seismic sections and have determined where they feel a prospective formation lies. If this is a brand-new prospect, a *wildcat well* is drilled—so named because it's " way out on the prairies where only the wildcats and hoot owls roam." This kind of drilling is high-risk, and the chances for finding a reservoir of commercial proportions are very low.

The other kind of drilling is called *development drilling.* In development drilling, usually little or no geophysical prospecting is done before drilling. The producing characteristics of the wells already completed nearby are available, and more is known about the subsurface structure.

With development drilling, the risk is considerably less than with exploration or wildcat drilling. Although the drilling operation is the same for both kinds of wells, development wells cost less than wildcat wells because more subsurface information exists. Geophysical prospecting usually is not needed, and less testing is done. Another important difference is that some of the expensive precautionary measures needed in wildcat drilling operations are not required in development drilling.

Table 4-1
Gray Energy Company
Well No. 1
Drilling Time Analysis

Item	Days	Percent of Total Time
Moving in and rigging up	2.5	6.3
Drilling	20.6	52.3
Down time for repairs	0.7	1.9
Formation evaluation		
Logging	0.6	1.7
Coring	1.1	2.7
Drillstem testing	0.8	1.9
Fishing0.9	2.2	
Cementing	1.5	3.8
Standing cemented	1.7	4.2
Conditioning mud	2.4	6.1
Well completion	5.0	12.8
Making decisions	0.5	1.3
Other	1.1	2.8
Total	39.4	100.0

Total depth = 9,550 ft
Well spudded March 26, 1988
Well completed June 20, 1988

In a very large company, the drilling section is usually part of the production department. Each operating area may have one or more drilling foremen to supervise both company and contract rig operations. Higher up, a drilling superintendent and a staff of engineers may answer to the production manager. Some large companies have no drilling superintendents but instead have a staff of drilling engineers reporting directly to the production superintendent, who is in charge of both producing and drilling operations.

The organization structure of an independent drilling contractor is relatively simple, since the contractor only drills wells—nothing else. The rig toolpushers report to the contractors' drilling superintendents, who supervise all rigs in a given locale. Contract toolpushers also report directly to the operator's drilling foreman. The drilling superintendents in turn report to the vice-president or president in the main office. Engineering and clerical staff are usually located in the main office, although some contractors have branch offices that have support personnel.

Of the two kinds of operations—in-house or contract—more than 90% of all wells in the Free World are drilled by independent contractors. There are many reasons for this, and the following top the list:

- The independent contractor, in business solely to drill wells, can usually do the job more economically than the producing company.
- Most companies drill only a few wells in any given locality. If they had to transport their rigs from one region to another, considerable expenses would be incurred.
- Like rig equipment, crews would have to be moved from location to location. That would mean higher wages and overhead.
- Company rigs need continuous drilling sequences, while contract rigs can be released.

All of these reasons contribute to the popularity of independent drilling contractors.

When a company uses an independent drilling contractor, a formal written agreement or drilling contract is drawn up. This contract sets forth the duties and responsibilities of the drilling contractor and the operator—duties that have been more or less established by custom. Like any contract, though, any provisions may be included as long as both parties agree. Over the years, however, three standard types of contracts have evolved: (1) the footage-rate contract, (2) the day-rate contract, and (3) the turnkey contract. All of these specify the basis on which the contractor will charge for drilling the well.

Under a footage-rate contract, which is the most widely used type, the contractor agrees to drill a well to a certain depth for a specified amount of money per foot drilled. Although the contractor is paid on a footage basis, payment is usually contingent upon drilling to the desired depth. If something prevents this—something not the fault of the operator—the contractor usually is not entitled to any pay for the footage he has drilled. Certain types of work, such as coring, logging, testing, and running casing, are considered extra work, and the operator usually compensates the contractor for these services on a day-rate basis.

Under a turnkey contract, which has become increasingly popular, the contractor furnishes rig, crew, and all equipment and materials needed to drill the well, including logging and formation testing equipment, drilling mud, and sometimes even casing and producing equipment. The operator assumes no responsibility other than to pay for the job. The contractor must deliver a properly completed well or a properly plugged dry hole before any payment is made. This contract is good because it gives the drilling contractor greater flexibility to ensure efficient, economical drilling. But because of the added risk for the contractor, this kind of contract usually costs more than a day-rate or even a footage-rate contract. A modified day-rate contract, used extensively in the Middle East, uses a basic day-rate basis at the contractor's cost plus footage rate profit margin as an incentive. This is considered fair for both parties.

When the contract has been settled on, all leases have been signed, and the proper permits have been attained from local, state, and federal agencies, drilling operations begin. The next step is development.

FIELD DEVELOPMENT

Once a commercially productive well has been drilled, proving the existence of an oil field, interest turns to the problem of determining how large the field is. The productive area must be determined, and the more productive sections must be located.

Once a given area is designated as productive, the operator within that area begins planning a development program that will protect his property from drainage from neighboring operators and that will provide the maximum economic return to the owner. One way the operator does this is to be sure the landman has done his homework and has leased as many adjacent properties as possible. Usually, though, many different operators own portions of a field and compete for production. Therefore, the location of initial wells in undeveloped areas is influenced by property lines as well as geologic structures. Protecting these property lines is a great consideration. More about this shortly.

Even though competent geological advice may be available, the early period of development of a field is uncertain. If a number of operators are competing with each other, the emphasis will be on speed of drilling to develop early production instead of securing accurate well data to help correlate and interpret the structure. Many operators consider their well logs

Fig. 4-2
The three types of reservoir drives.

confidential information. In these cases, it becomes difficult for operators to determine the structural and stratigraphic relationships—information fundamental in planning a development program.

When planning a program, several characteristics are very important to remember. We discussed in Chapter 3 the importance of water and gas drives (Fig. 4-2). When the natural reservoir energy is utilized to its full potential, the reservoir will produce more efficiently and economically. Early wells experience higher reservoir pressure and greater production time, so their initial and ultimate production is substantially greater than wells drilled later. A delay of even a few months may mean substantial losses in ultimate production.

Another important aspect of timeliness deals with obtaining more oil than a neighboring property. The operator who first brings this property to full development can secure more of his neighbor's oil. Oil and gas migration doesn't stop at the property line; they move to the nearest well. Theoretically, the first wells drilled have greater ultimate productions if completed in the best part of the section. Initial production will be greater again because of the higher gas pressure during the early stages of development. So the "first come, first served" policy is an important factor. It all boils down to timeliness.

Information gained from drilling the first well is integrated with previous information (logs, maps, and pressure and production data) to determine the extent of the field and its estimated recoverable reserves of oil and gas. Then a development plan is formulated, which considers the total number of wells required, the spacing or distance between the wells, and the geometric pattern for placing the wells.

Estimating the Number of Wells

Usually an operator wants to develop the largest possible area with the fewest number of wells without running the risk of locating a well beyond the limits of the accumulation and drilling a dry hole. Well productivity must also be considered and weighed against the cost of drilling additional wells to determine the ideal number of wells that will produce the greatest rate of return. To make the determination, a skilled petroleum engineer applies economic principles combined with technical talents to develop an optimum plan.

To reduce risk, the second well is generally a step-out well—a well located only "a step away" from the first well, not a great distance away. The reservoir engineer combines geological and drilling information to determine from the information gained from the initial well the most favorable direction for further development from the discovery well. The type of structure and the magnitude and extent of the trap or fold are important considerations in determining the position of the second, third, and later test wells and the distance that they may be spaced from the first well.

Fig. 4-3
During the early years, no guidelines were established for well spacing. At Spindletop, you could step from one rig to another.

Spacing

Some geometric arrangement is usually followed in placing wells. Often the spacing and arrangement of boundary wells will determine the position of interior wells, particularly if the property is small. There is more opportunity for scientific well spacing and arrangement when land is held in large tracts rather than small acreage.

If the structure indicates a well-developed anticline or dome, exploration for the limits of the productive area may be conducted by drilling wells first in both directions along the major axis of the structure, locating the wells as closely as possible along the structural crest and, second, along a line at right angles to the axis. The wells are located alternately on either side of the crest, exploring down the flanks until edge water (water surrounding the productive formation) is encountered or until the wells become such small producers that they cease to be profitable.

Well spacing is not a matter than can be decided easily. The physical and economic conditions in each situation should be considered carefully before decisions are reached. The expense for drilling wells should be weighed against the profits obtained to determine the most economic combination. The operator wants to determine the number of wells required to produce the greatest profits. Because of the great number of complex variables, this often becomes a trial-and-error situation.

Well Pattern

Some geometric arrangement is usually followed in placing wells. Often, the spacing and arrangement of boundary wells will determine the position of interior wells, particularly if the property is small (Fig. 4-4). There is more opportunity for scientific well spacing and arrangement when land is held in large tracts rather than small acreages.

In essence, spacing involves drilling preliminary, widely scattered wells at some uniform distance. After this primary system is completed, intermediate wells are drilled between the previous wells at an interval and spacing designed to produce the most economic extraction. This plan has three advantages:

- Initial production is higher than production using plans with usual spacing.
- Production from widely spaced wells is sustained better than closely spaced wells.
- The decision on ultimate spacing can be deferred until more information is available.

The obvious disadvantage is, of course, the risk of drilling a dry hole. All such risks must be weighed before a decision is made.

The development of an oil property is generally conducted according to one of several plans. A common method is to drill rows of wells across the property from proven territory to unproven territory. This plan gives maximum protection from the risk of drilling a dry hole when the operator is not certain that the entire area beneath the property is productive. It also offers the opportunity to secure vital information on structural and sub-surface conditions for new locations before drilling the next location. A somewhat similar plan is to drill progressively outward from productive test wells as centers.

In the U.S., the general rule is one well per 40 acres. In Canada, well spacing is somewhat less dense. And in the Middle East, where individual well productivity is high, typical spacing can be as broad as one well per 640 acres. It all depends on the government and its individual regulations.

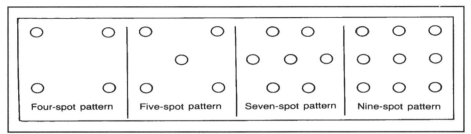

Fig. 4-4
Some typical well patterns.

After the number of wells, the spacing distance, and the pattern are chosen, the wells are drilled. As wells are drilled, they are numbered on each property in the order they are drilled. Large companies sometimes number the wells with reference to their position, regardless of the order of drilling. This can be advantageous because it identifies location immediately; the disadvantage is that you can't readily tell when the well was drilled.

Additional Considerations

When planning a development pattern, other factors come into play too:
- Reservoir drive.
- Controlling rate of production.
- Amount of surface equipment needed.
- Proximity to utilities and carriers.
- Market outlet and value.
- Government regulations.

Each of these factors must be considered in the overall plan.

We discussed earlier the three types of drives: water drive, dissolved-gas drive, and gas-cap drive. Depending on the contents of the reservoir and the depth to which a well is drilled, the well may or may not be developed to its most productive capacity. If the driller does not take advantage of the natural energy in the reservoir, some kind of lift or pump will be needed to produce the petroleum, and that is an additional charge. So knowing the type of drive is important.

Controlling the rate of production is also important. Efficient recovery is not obtained by chance; it is accomplished only by careful and deliberate action on the part of the operator (Fig. 4-5). Experience has proved that one of the most essential factors in meeting the requirements for efficient oil recovery is controlling the rate of production. Excessive rates lead to rapid decline in reservoir pressure, premature release of dissolved gas, irregular movement of the gas or water displacement fronts, dissipation of gas and water, trapping and bypassing of oil, and in extreme cases dominating the entire recovery with an inefficient dissolve-gas drive. Each of these effects, resulting from excessive withdrawal rates, reduces the ultimate recovery of oil. Generally, operators recognize the most effective method of controlling the displacement mechanism for increased ultimate oil recovery is to restrict the oil production rate.

The kinds of necessary surface equipment are also important in the development plan (Fig. 4-6). As soon as the first well is drilled and placed on production, oil storage tanks and gathering and treating facilities for both the oil and gas will be needed. This portion of the surface equipment is ordinarily developed gradually to keep pace with the productivity of the wells as they are completed. However, the layout and design must be carefully planned in

Fig. 4-5
Lake View No. 1 well, known as the Lakeview Gusher in Kern County, Calif., blew out while drilling on Mar. 10, 1910, and flowed uncontrolled at rates of as much as 68,000 b/d until September 1911, when it ceased flowing, apparently because of bottomhole caving (courtesy Kern County Museum).

Fig. 4-6
A simplified surface equipment setup. Note the separators (1), treater (2), disposal tank (3), and stock tanks (4).

advance to accommodate the shape and size of the property and the typography.

The proximity to utilities and carriers is another consideration. If the property is in a remote location, roads may have to be built, electrical wires may have to be strung, and temporary housing may have to be provided for crewmembers. If a pipeline is not close by in which to transport gas, a line must be built or the well must be capped until the field shows greater promise.

Closely aligned to this is market outlet and market value. Both of these considerations depend on timeliness. And when time is required to provide pipeline connections and to negotiate sales contracts, big fluctuations can occur in the price for a barrel of oil or a thousand cubic feet of gas (Mcf). Drilling expenditures present great financial hazards until the potential of a prospective area is determined. Financial institutions generally will not lend capital for drilling wells because of the risk involved. So, many operators are forced to adopt a conservative drilling program in which profits on early wells contribute capital for subsequent drilling.

Perhaps the biggest consideration of all—certainly one that changes often—is government regulation. In the United States, Canada, and most other countries, permission must be secured from the appropriate governmental agency, either state or national, to drill an oil or gas well. In many areas, certain minimum spacing requirements have been established. Unless special permission is obtained, a well cannot be drilled before the operator has assembled the required number of acres.

All of these factors affect field development. The entire process is much more involved than just constructing a rig and drilling a hole in the ground. It's a combination of big business, high risk, and the cutting edge of scientific know-how—all working together to squeeze that oil or gas out of the ground and bring it to the surface.

CHAPTER 5

Drilling Equipment and Methods

Drilling and production methods really cannot be separated into two categories. Perhaps more than any other two functions in oil well technology, drilling and production are inseparable. Because of this, let's take a quick look at some of the basic fundamentals of drilling before we investigate how a formation is evaluated.

TYPES OF RIGS

The most common type of land drilling rig is the *cantilevered mast*. This is assembled on the ground and then is raised to the vertical position using power from the rig's drawworks hoisting system (Fig. 5-1). Sometimes this is called a *jackknife* derrick rig.

When assembling this rig, the outside structure, made of prefabricated sections, is joined together with large pins. The engine and derrick sections are put into position and pinned together by the drilling crew. Next, the drawworks and engines are put into place. Finally, the derrick sections are laid out horizontally and raised as a unit by the hoisting line, traveling block and drawworks.

Offshore rigs perform the same function, but their design is more complex (Fig. 5-2). In shallow waters or swamps, a *barge* is used. A barge is a shallow-draft, flat-bottomed vessel equipped with a jackknife derrick. A *jack up rig* operates in water as deep as 350 ft. These rigs are very stable because they rest on the seafloor. The rig's jacket is slowly towed to the location during calm seas. Then the legs are lowered by jacks until they rest on the seafloor below the deck. The legs continue to lower until the deck is lifted off the surface of the water (sometimes 60 ft) and the deck is level.

Fixed-platform rigs, yet one more of many kinds of offshore vessels, are pinned to the seafloor by long steel pilings. These platforms are very stable but

Fig. 5-1
Jackknife cantilevered mast, used on a rotary rig.

are considered permanent and virtually immobile. (The jack up, on the other hand, can be moved from location to location.) *Semisubmersible rigs* can also operate in water 200-500 meters deep, and they are equally stable but not fixed. These rectangular floating rigs carry a number of vertical stabilizing columns and support a deck fitted with a derrick and related equipment. But the rig that offers the greatest mobility and that can operate in almost any water depth is the *drillship*, a ship specially constructed or converted for deep-

Fig. 5-2
Types of offshore drilling rigs.

water drilling. Dynamic positioning equipment keeps the ship above the well-bore using a thruster with controllable pitch propellers.

DRILLING METHODS

Whether drilling on land or offshore, a successful drilling system must provide (1) a means of fracturing or abrading the rocky formations that must be penetrated to reach the oil or gas; (2) a means of excavating the loosened material from the well as drilling proceeds; and (3) a means of preventing the walls of the well from caving in as well as a way of sealing off water and gas. In addition, generally the wells need to remain as close to vertical as possible, the well must be deep enough to reach the reservoir, and the diameter must be large enough to permit lowering tools into the hole.

Today, the two most widely used drilling methods are cable-tool and rotary drilling. Although rotary drilling is used more frequently, the earliest method used was cable-tool drilling.

Cable-Tool Drilling

In the cable-tool method, drilling is accomplished by lowering a wire line or cable into the hole. On the end of the line is a heavy chisel-shaped piece of

Fig. 5-3
Diagram of cable tool drilling rig.

steel called the drilling bit. An up-and-down motion is applied to the line at the surface. This churning action chips small pieces of rock from the formations (Fig. 5-3).

The bit, which weighs several hundred pounds, is continuously dropped until a few feet of hole have drilled. At this time, the line is raised by a drum at the surface and the bit is removed. Then a bailer, a metal pipe with a one-way flapper valve on the lower end, is lowered into the hole on another line called the sand line. The chips cut by the drilling bit are picked up in the bailer and removed from the hole so drilling can resume.

In cable-tool drilling, no significant amount of fluid is circulated in the hole. Ordinarily, the only fluid in the hole is what has come from the formation being penetrated. A small amount of water is desirable, however; if no water comes from the formations, a few gallons are dumped downhole.

One of the greatest advantages of the cable-tool is that it helps quickly identify producing oil and gas zones. It is also useful in drilling certain formations that are sensitive to water-base drilling fluids. In some formation, there is a chemical reaction between the water in the drilling fluid and the clays in the rocks. This results in a decreased rate of oil or gas flow. Use of the cable-tool method minimizes this problem because of the small amount of water used.

Drilling with a cable-tool rig is a slow process. Often a cable-tool rig drills only one-tenth as fast as a rotary rig in comparable formations. However, the cost of a cable-tool rig is substantially less than a rotary rig. This tends to compensate for its slower drilling rate.

A distinct disadvantage of the cable-tool method is that when high-pressure oil and gas formations are encountered, there is no fluid in the hole to control them. The result is frequent blowouts. When a *blowout* occurs, the oil and gas from the subsurface formation rush to the surface and flow uncontrolled. A blowout may spray the oil and gas several hundred feet into the air, and there is always great danger of a fire.

Because of its slow penetration rate and the hazard of blowout, the cable-tool method is seldom used on wells deeper than 3,000 ft. Even on shallower wells, this method has largely been replaced by the rotary method.

Rotary Drilling

Rotary drilling is quite different from cable-tool drilling. In rotary drilling, a bit used to cut the formation is attached to steel pipe called *drillpipe*. The bit is lowered to the bottom of the hole. The pipe is rotated from the surface by means of a rotary table, through which is inserted a square or hexagonal piece of pipe called a *kelly*. The kelly, connected to the drillpipe at the surface, passes through the rotary table. The turning action of the rotary table is applied to the kelly, which in turn rotates the drillpipe and the drilling bit.

1. Accumulator
2. A-frame
3. Air compressor
4. Annular (bag) preventer
5. Annulus
6. Base
7. Bell nipple
8. BOP control
9. Bit (drill)
10. Bradenhead
11. Burning pit
12. Casing-hanger spool
13. Cathead
14. Cat line
15. Catwalk
16. Cellar
17. Centrifuge
18. Chemical barrel
19. Choke line
20. Choke manifold
21. Choke manifold control
22. Compound
23. Conductor casing
24. Crown block
25. Cyclone desander desilter
26. Dead line
27. Degasser
28. Discharge line
29. Doghouse
30. Drawworks
31. Drill collars
32. Driller's console
33. Drilling line
34. Drillpipe
35. Drill tool storage (junk box)
36. Dynamatic or hydromatic
37. Elevators
38. Engines
39. Fast line
40. Fill-up line
41. Flow line
42. Fuel line
43. Fuel tank
44. Generating unit (light plant)
45. Gin pole
46. Hoisting line
47. Hook
48. Intermediate casing
49. Kelly
50. Kelly bushing
51. Kelly (rotary) hose

52. Kill line
53. Ladder
54. Line guide
55. Mast
56. Mast lifting line
57. Mixing (mud) pit
58. Monkey board
59. Mousehole
60. Mud
61. Mud-gas separator (gas buster)
62. Mud gun (submerged)
63. Mud gun (surface)
64. Mud hopper
65. Mud line
66. Mud logging unit
67. Mud (paddle) mixer
68. Mud-mixing plant
69. Oil and grease storage
70. Pipe rack (floor)
71. Pipe racks
72. Pressure (mud) gauge
73. Preventer control lines
74. Preventer (BOP) ram type
75. Production casing
76. Pump drive
77. Pump, mud mixing
78. Pumps, mud
79. Ram wheel
80. Ramp
81. Rathole
82. Reserve drilling line
83. Reserve (mud) pit

Detail for 113

Fig. 5-4
Drilling rig schematic.

65

Driller's side

Back

Front (V-door)

Mud-pit side

84. Rotary drive
85. Rotary (table)
86. Safety (Geronimo) line
87. Sand settling (mud) pit
88. Shale pit
89. Shale pit jet
90. Shale shaker
91. Stairs
92. Stand pipe
93. Storage, bulk barite
94. Storage, liquid mud
95. Storage, mud additives

96. Substructure
97. Substructure, pony
98. Suction lines
99. Suction (mud) pit
100. Surface casing
101. Swivel
102. Swivel (spring)
103. Tong, counterweight
104. Tong, lead (breakout)
105. Tong, makeup

106. Tool house
107. Traveling block
108. Trip tank
109. Walkway
110. Water, storage
111. Water table
112. Weight indicator
113. BOP and casing

Fig. 5-4
Drilling rig schematic (continued).

Routine drilling consists of continuously drilling increments the length of one joint of pipe, making connections or adding to the drillstring another single joint of pipe, generally 30 or 45 ft long. This drilling continues until the drill bit must be changed—when it is worn or when a formation is penetrated for which the bit being used is not well suited. Changing the bit is also called *making a trip*. A *round trip* is simply coming out of the hole, changing the bit, and going back into the hole.

Three principal types of bits are used in a rotary drilling operation: (1) drag or fish-tail bits, (2) rolling-cutter bits, more commonly called rock bits, and (3) diamond bits. Most drilling bits are rock bits, of which there are many different styles and types for cutting various formations.

A drilling rig consists of many components, each of which has an important function. The principal components of a rotary rig are (1) the mast, (2) the drawworks, (3) the engines, (4) the mud system, and (5) the drillstring (Fig. 5-5). The mast or *derrick* is the structure placed over the well the help remove the pipe from the lower equipment into the hole. The drawworks is the hoisting equipment. The engines drive the mud pumps and drawworks and provide power for miscellaneous requirements like electricity. The mud system is comprised of the mud pumps, the mud tanks, the mud flow lines, and the circulating hose. The drillstring is the entire rotating assembly and consists of the kelly, drillpipe, drill collars, and drill bit.

At the bottom of the hole, the cuttings, or pieces of formation cut loose by the drilling bit, are removed from the hole continuously through the circulation of drilling mud or fluid. This mud or fluid is circulated down through the inside of the drillpipe and up again, outside the drillpipe, to the surface—the primary function of the drilling mud. However, drilling mud has other important functions: it cools and lubricates the bit and plasters the wall of the hole, making the hole more stable; and its hydrostatic pressure prevents the flow of salt water, oil, and gas into the wellbore, helping prevent blowouts.

DRILLING FLUIDS

Of course, only a few gallons of water are used as fluid in cable-tool drilling. But in rotary drilling, fluids are very important.

The most common type of fluid, or mud, is a suspension of clay in ordinary fresh water. This clay, also called bentonite or gel, is mixed in a finely dispersed form which results in a relatively smooth, homogenous mixture. Often, chemicals must be added to this clay-water mixture to improve its properties.

An important part of most drilling fluids is weighting material, of which the most common is barite. Weighting material increases the density of the drilling mud to help overcome high-pressure flows of oil, gas or salt water. To eliminate certain drilling problems, special types of clay-water muds are used.

Fig. 5-5
Components of a rotary drilling rig: (a) mast, (b) drawworks, (c) mud-system engines, (d) mud system and drillstring.

Another type of drilling fluid commonly implemented is oil-base or oil emulsion mud. Oil-base muds are used where formations would be damaged by water-base drilling fluids. These muds are more expensive than water-base drilling fluids.

The use of air or gas as a drilling fluid is increasing. In certain situations, air and gas may have many advantages over the water- or oil-base drilling muds. One principal advantage is that air or gas will not damage the formation as water-base muds often do. Another advantage is that penetration rates are substantially greater with air or gas as the drilling fluid.

Foam or mist drilling is a special type of air drilling in which small amounts of water and chemical are added to the air as it is pumped downhole to form a foam or mist. Foam or mist drilling is used where large volumes of water are coming into the wellbore from the formations being drilled.

This, of course, is a very simplified overview of drilling operations. However, let's continue now and look at how a formation is evaluated to determine if it contains commercially producible quantities of hydrocarbons.

CHAPTER 6

Formation Evaluation Logging, Coring, and Drillstem Testing

At this time, no instrument conclusively indicates the presence of oil underground. Geologists and geophysicists can suggest the most probably geographical location and geological time periods in which oil is likely to be found in significant accumulations. But only until an exploratory well is drilled can engineers have a peek into the underground formations the drill bit has penetrated. In exploratory drilling, it is necessary to have a group of methods and tools to locate and evaluate the commercial significance of the rocks the drill bit penetrates. We call the use and interpretation of these methods *formation evaluation.*

Formation evaluation methods can be broadly classified according to whether they are used (1) as drilling is in progress or (2) after the well (or at least a portion of it) has been drilled. In the first classification are drilling fluid and cuttings analysis logging, and coring and core analysis. In the second classification are wireline logging, sidewall coring, wireline formation testing, and drillstem testing.

Of the many methods available for formation evaluation, no one is of any great value on its own. Each complements another.

WELL LOGS

A well log is any tabular or graphical portrayal of well conditions. Some of the well logs most widely used for formation evaluation are listed below:
* Mud logs.
* Pressure logs.
* Core logs.
* Wireline logs.

In addition to these kinds of logs, engineers use two other kinds of logs to help determine characteristics of formations. These logs—the driller's log and the sample log—are very important records and, because they are monitored constantly, not just during drilling breaks, can often indicate promising formation. Then, one of the more sophisticated and costly logs can be run to verify the initial observations and to help determine how productive a formation will be. Let's first look at the driller's log and the sample log.

Driller's Log

The driller is in charge of the rig and crew for an 8 or 12-hour shift, called a "tour" (pronounced "tower"). Each driller prepares a driller's log, which is a record of the operations and progress during his tour (Fig. 6-1). The log contains geological descriptions and mechanical information for each well. It may also indicate if any pertinent fluid flows are encountered or if any "shows" of oil or gas were observed.

Included along with the driller's log is the driller's time log. This log is often kept when the well approaches a zone of particular interest. It may also be kept continuously when a well is being drilled in an area where little is known about the depths of potential zones that may contain hydrocarbons. Penetration rates are important because hydrocarbon-bearing formations are generally softer and drill at a faster rate than the hard rock formations often encountered above or below. *Drilling breaks*, or changes in penetration rates, can sometimes be quite dramatic, while other times they may occur more subtly.

Sample Logs

As the drill bit makes its way through the earth, penetrating rocks, the bit produces *cuttings*—small pieces of pulverized rock chips. The cuttings are circulated back up to the surface by the drilling fluid. These cuttings are a source of significant information if they are properly procured and evaluated by a competent geologist.

In and exploratory well, cuttings samples may be taken at regular intervals from the entire wellbore. In a field development well, this coverage is generally unnecessary and only the zone(s) of interest may be sampled.

In rotary drilling, part of the drilling mud stream is diverted into a sample box. The rate of the stream is slowed, and the samples filter down to the bottom of the box. At a designated interval, the samples are removed, washed, placed in a bag, and marked—ready for a geologist to inspect them.

Care must be taken in collecting samples. If the time is not recorded accurately, the right depth won't be matched, and a formation could be missed. There is also a *lag time* difference—the amount of time it takes for the cuttings to circulate from the bottom of the hole back to the surface. In deep wells, this

SPECIMEN WELL-LOG RECORD
DRILLERS LOG
(Front Side.)

Field Coalinga

COMPANY
California Oilfields, Limited
(Shell Co. of California)

Log of Well No. 78
Description of Property (Quarter Section) S.W. ¼ of Sec. 27, 19-15
Location of Well 740' N. and 2905' W. of S.E. corner
Elevation Above Sea Level 1,178 ft.
Commenced Drilling Oct. 29, 1913. Finished Drilling

Depth from—	To—	Feet.	Formation
0	10	10	Brown Adobe.
10	25	15	Brown sand.
25	55	30	Yellow clay.
55	65	10	Coarse gray sand.
65	98	33	Black gravel.
98	125	27	Brown sand.
125	185	60	Blue sandy shale.
185	210	25	Light blue shale.
210	245	35	Coarse gray sand.
245	315	70	Light blue sale.
315	317	2	Brown shale.
317	330	13	Blue shale.
330	390	60	Sandy blue shale.
390	404	14	Fine gray sand.
404	440	36	Light green shale.
440	450	10	Gray sand.
450	478	28	Coarse gray sand and gravel.
478	497	19	Gray sandy shale.
497	510	13	Coarse gray sand.
510	535	25	Sandy blue shale.
535	580	45	Blue shale.
580	640	60	Sandy blue shale.
640	690	50	Gray sand, shows tar oil.
690	705	15	Blue shale.
705	715	10	Sandy blue shale.
715	723	8	Gray sand, shows tar oil.
723	733	10	Fine hard gray sand.
733	740	7	Hard sand shell.
740	753	13	Gray sand, shows tar oil.
753	757	4	Blue sand shell.
757	785	29	Soft gray sand.
785	796	11	Blue shale.
796	797	1	Hard sand shell.
797	805	9	Soft sand and gravel Water. (Water stands at 600'.)
805	806	1	Hard sand shell.
806	870	64	Sticky blue shale.
870	905	35	Fine gray sand.
905	920	15	White sand and sea shells. (Put in 2 loads red mud at about 930'.)
920	965	45	Soft gray sand.
965	985	20	Sandy blue shale.
985	1,005	20	Sandy shale, black.
1,005	1,055	50	Fine soft gray sand.
1,055	1,092	37	Hard coarse gray sand.
1,092	1,104	12	Sticky black shale.
1,104	1,129	25	Sticky light blue shale.
1,129	1,140	11	Light gray slate.
1,140	1,214	74	Tough green snale. (12½" casing cemented at 1214'.)
1,214	1,233	18	Tough, sticky green shale.
1,232	1,280	48	Light green shale.
1,280	1,295	15	Light blue shale.
1,295	1,305	10	Light gray shell.
1,305	1,330	25	Sticky blue shale.
1,330	1,348	18	HARD GRAY OIL SAND, fair.
1,348	1,363	15	FINE GRAY OIL SAND, good.
1,363	1,380	17	Hard gray sand, no oil.
1,380	1,393	13	SOFT GRAY OIL SAND.
1,393	1,410	17	Hard gray sand, no oil.
1,410	1,421	11	Black sandy shale
1,421	1,423	2	Hard sand shell.
1,423	1,440	17	Fine black sand.
1,440	1,445	5	Hard sand shell.
1,445	1,470	25	Fine dark gray sand.
1,470	1,493	23	Sandy blue shale. (10" casing cemented at 1626'.)
1,493	1,495	2	Hard sand shell.
1,495	1,500	5	Very sandy shale, shows oil and gas.
1,500	1,510	10	Soft fine gray sand, shows oil.
1,510	1,525	15	Light blue shale.
1,525	1,587	62	Gray sand, shows oil and gas.
1,587	1,598	11	Black sandy shale.
1,598	1,608	10	Hard fine gray sand, no oil.
1,608	1,620	12	Fine black sand, shows Sulfur Water.
1,620	1,629	9	Tough black shale.

Fig. 6-1
Sample driller's log.

lag time can often be hours, so the sample collector must be precise in recording and labeling his samples.

What do samples show? Several things:

- Rock type (sandstone, shale, limestone, etc.).
- Specific formation penetrated.
- Age of the rock.
- Well depth at which formation was encountered.
- Indications of porosity, permeability, and oil content.

These data, plus information from the driller's log, are some of the easiest measurements to make during drilling (see Fig. 6-2). In addition to using these kinds of methods, logging specialists may be called in to run more specific kinds of logs on the well. These kinds of logs include the mud log, the pressure log, core logs, and wireline logs.

Mud Logs

Mud logging is the continual inspection of the drilling fluid and cuttings for traces of oil and gas. In part, it serves as a primary lead to coring and testing. It has an added usefulness as a safety measure for the early detection of hazardous drilling conditions that could result in loss of well control or blowout. Most services also inspect, analyze, and describe cuttings.

Generally, a contractor with a portable lab analyzes the drilling fluid returns for hydrocarbon content. A technician, usually a geologist, operates the laboratory, preparing the log and providing the operator with current information or drilling progress (Fig. 6-2). If shows or oil or gas are detected, the driller and technician know a hydrocarbon-bearing formation has probably been penetrated.

So what is the difference between a sample log and a mud log? A sample log is prepared by the geologist after he or she analyzes the cuttings samples. A mud log is an enhancement of the sample log based on continuous analysis of the drilling mud, looking for small traces of hydrocarbons in the fluid.

There are two general methods of using mud logs. A primary purpose is to evaluate the formation as the well is being drilled to control points for casing, coring, testing, or further evaluation. If this is the purpose of the log, drilling proceeds, and the hole is logged without interruption. When there is a significant increase in drilling rate, or when other occurrences such as an increase in total gas and/or the appearance of heavy gases indicates a possible reservoir has been penetrated, then drilling ceases.

In this second case, after about 10-15 ft of penetration, drilling is halted, and the cuttings from the break are lifted to the surface. If no show is indicated, drilling may be resumed with a minimal loss of time. However, if the cuttings indicate possibilities of oil and gas, a core may be cut or a drillstem test may be considered. More about cores and drillstem tests later.

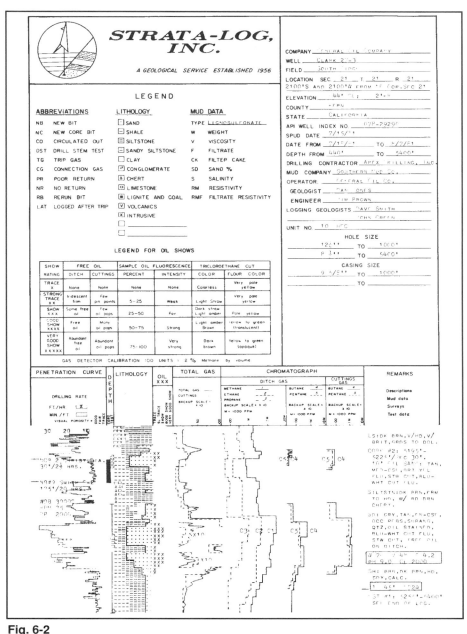

Fig. 6-2
A mud log. Note the lithology column, which is used for recording characteristics of samples (courtesy Strata-Log).

Mud logs are generally applied to exploratory wells and field development wells where specific problems exist. These might be exploratory wells in areas where detailed subsurface information is lacking; field development wells in areas where lensing sands, folding, and faulting make subsurface correlation difficult; wells that are expected to encounter high-pressure formations; and well in areas where electric log interpretation is difficult.

Several kinds of information are furnished by the mud log:

- Direct measurement of hydrocarbon gases from the drilling mud.
- Chromatographic analysis of the drilling mud for individual hydrocarbon gas content.
- Total combustible gas from drill cuttings.
- Oil from drilling mud and cuttings.
- Detailed rate of penetration curve.
- Lithology log and description with estimated porosity.
- Drilling mud characteristics.
- Data pertinent to the well's operations (e.g., trips for a new bit).
- Bit data, carbide information, deviations, and other pertinent engineering information.

In addition, mud logging techniques have many advantages:

- The results are available almost immediately.
- The procedure does not interfere with drilling operations.
- The log is recorded simultaneously with the driller's log.
- Detailed data on subsurface characteristics are collected continuously and analyzed at the surface.

In addition to almost immediately indicating the presence of any potentially productive zone, the mud log serves as a basis for tailoring and altering the drilling program efficiently. It is an important corroborative and correlative tool.

Pressure Log

The pressure log is a computer analysis of certain drilling parameters and data (Fig. 6-3). Information from a number of sources at the well site is continuously monitored by a computer that produces a constant estimate of formation pressures. It is normally used on exploratory wells or in areas where pressures are difficult to predict.

Abnormal formation pressure can be evaluated graphically by recording the gas shows, their magnitude, character, and behavior; associating these with other factors; then linking all factors with geological indicators like formation type and size of cuttings. Pressure is an important measurement because it relates to porosity. Formations exhibiting high pressures at a particular depth are usually zones of abnormally high porosity. The high-porosity abnormality is based on the principle that, as burial depth increases, overburden pressure increases, thereby compacting the rock bulk.

A number of pieces of specialized equipment with surface sensors evaluate the downhole pressure while drilling. These systems, in addition to gas monitors, include the following:

- Continuous mud weight, temperature, and resistivity recording.
- Bulk density and shale factor determination kits.
- Pit volume totalizer.
- Differential mud flow measurement.
- Computational hardware and software.

Core Log

The core log is a record of core analysis and lithology versus depth. Core analysis is used for exploring and evaluating the productive possibilities of edge wells and exploratory wells. In the development of a field, core data guide the driller when completing wells. Cores also yield information for the preliminary evaluation of the oil property. Finally, they are used to engineer and design enhanced recovery (EOR) methods for the field. Good core data such

Fig. 6-3
Portion of pressure log (courtesy Exlog).

as porosity, permeability, and fluid saturation are necessary for good reservoir operations and prediction of performance. We'll learn more about cores and how they are obtained and analyzed in the section on coring.

Wireline Logs

One of the largest categories of logs are those offered by specialized companies that record and measure the signals that are traced or emitted by various tools lowered downhole on wire cable. Commonly known as wireline logs, these tools provide data that is very important in formation evaluation.

Wireline logs are run by lowering a sonde and cartridge into the hole and then pulling it up at a fixed speed, determined by the measurement to be made (Fig. 6-4). As the tool is withdrawn from the hole, a continuous measurement signal is sent to the surface through the conductors in the cable. The raw data is processed in a control panel and recorded in the proper log format on film by an optical recorder.

Fig. 6-4
Running a well log sonde.

These logs are generally classified as either electrical or radioactive logs, although there are several kinds of logs that for the purposes of this book we categorize as "miscellaneous." Let's take a look at some of these logs and what they can do in the formation evaluation process.

Electrical Logs

The electrical log, one of the most widely used today, is obtained by lowering an instrument into the wellbore on an insulated electric cable after the drillpipe has been removed from the hole. Each type of formation has a characteristic electrical response, and oil and gas respond differently than water. The electric log measures the electrical properties for the formations and their fluids. Thus, with proper interpretation the log can indicate whether a formation contains oil and gas as well as the nature of the formation (sandstone, limestone, or shale).

Electric logs are called *open-hole logs* because they cannot be run in cased holes. The steel pipe interferes with the electrical responses of the formation.

When the logs are recorded, several kinds of information are plotted (Fig. 6-5). The standard electric log will record two sets of graphs. The left-hand side presents the spontaneous or self-potential (SP) curve, and the right-hand side presents the resistivity measurements. Several different measurements may be recorded simultaneously with each run of the instrument.

The spontaneous potential is the small, minute amounts of electrical voltage that almost all materials exhibit in varying amounts of degrees. Voltage is to electricity as pressure is to a fluid: both represent the potential for something to try to flow. In the case of electricity, it is the potential or pressure for electrons to flow.

Electrical resistivity is the measure of materials resistance to the flow of electrons. It may be thought of as the opposite of conductivity, the ability or property of a material to conduct electricity or the flow of electrons. Resistivity gives important clues of a formation's probable lithology and fluid content.

Thus, both the spontaneous potential and the resistivity of a formation give the geologist and engineer important clues about a formation's potential producibility.

There are numerous adaptations and configurations of electrical logging devices. A *laterolog* forces current to flow radially through the formation in a sheet of predetermined thickness by means of an appropriate electrode arrangement and an automatic control system. The measured value is unaffected by the drilling fluid in the well. A *microlog* is essentially a resistivity log with electrodes mounted at short spacing in the face of a rubber pad. The insulating pad is positioned against the wall of the borehole, reducing the short-circuiting action of the drilling fluid. This log measures only a small volume of material in

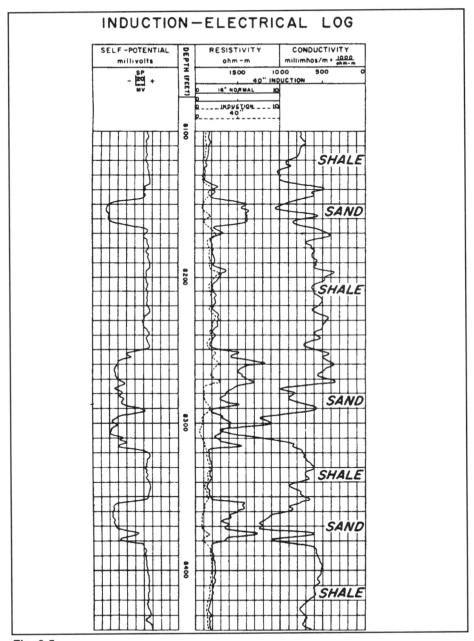

Fig. 6-5
An electric log of a well. Note the SP curve on the left and the resistivity curve on the right.

front of the pad, which is useful in recording hole diameter and the presence of mud cake. The *microlaterolog* has one center electrode and three circular ring electrodes around the center electrode. Current is applied to the electrodes, and the voltage is measured and recorded as a measure of resistivity.

These, of course, are very brief descriptions of electrical logs. Other devices are in current use, but these are some of the better known. Logging service companies, eager to compete, continually develop new tools that may give better and more accurate information on the nature of rock formations and the fluids they contain.

Radioactive Logs

Electric logs must be run in an open hole to avoid short-circuiting through the steel casing. But radioactive logs may be run in either open or cased holes.

Two curves are presented in a complete radioactive log: the gamma ray curve and the neutron curve (Fig. 6-6). The gamma ray curve is presented on the left-hand side and is similar to the SP curve. The neutron curve appears on the right-hand side and is like the resistivity curve. Together, these curves indicate natural and artificially produced radiation within a well.

A *gamma ray logging device* generally consists of an ionization chamber charged with an inert gas under high pressure. Gamma radiation emitted form the rock formations penetrates the ionization chamber. Some of the rays collide with gas atoms, liberating ionized electrons from the gas and thereby generating a current that is amplified at the surface and recorded relative to depth. The magnitude of the current is directly related to the intensity of the gamma radiation.

Since the gamma ray curve or log measures the natural radioactivity of the formations, the radiation varies with the type of rock. Rocks containing shale have the greatest radioactivity and are measured by right-hand deflections on the log. Igneous rocks are more radioactive than sediments and can be identified easily by radioactive logs.

The *neutron log* is obtained by moving a source that emits energetic neutrons along the wellbore together with a radiation detector at a fixed distance from the source. The source bombards the rocks with a constant flow of neutrons, and the detector records the varying intensity of the gamma radiation. The resultant curve is a measure of the fluid contained in the rocks.

Neutrons are emitted into the wellbore at high velocity, uniformly, in all directions. As they travel outward, they are scattered and slowed by collisions and finally captured. The properties of the surrounding materials are such that in well logging, the range of most of the surrounding neutrons is never as great as the source detector spacing. The response of the detector increases as the hydrogen content of the surroundings diminishes.

Sand or limestone
Shale
Sand or limestone
Shale

Sandstone or lime

Volcanic ash

Shale

Anhydrite

Shale

Shale

Shale

Shale

Limestone

Granite

Gamma ray   Neutron

Fluid bearing

Dense   Non-fluid bearing
Shale
Fluid and dense sand
Shaly sand or lime

Dense   Fluid and dense

sand or lime

Caprock
Salt
Anhydrite
Shale

Anhydrite

Shaly sand grading
to clean sand

Sandstone
Limestone

Anhydrite

Limestone

Granite

Radiation intensity increases

Fig. 6-6
Radioactivity log.

Miscellaneous logging devices

There are many other different types of logging devices that have varied success. Some of the more important are listed in this section.

The *acoustic log* uses ultrasonic signals that are generated and pass through the drilling fluid to the adjacent rock formations. The resultant signal is refracted parallel to the borehole and then is picked up by the receiver. The velocity of the signal that travels through the rock formations is measured by the logging tool and is recorded at the surface.

The information from acoustic logs can provide input on the lithology of the formations. Since different rock types exhibit different sound velocities, the rock types are easily differentiated from each other. Porosity values can be estimated since sound travels more slowly in oil and gas than in water; fluid saturations also can be estimated from travel time measurements.

The *caliper log* is a record of hole size diameter versus depth. This tool consists of a set of springs that expand against the wellbore (Fig. 6-7). A center rod is connected to the lower ends of the springs, which telescope into a cavity containing an electrical coil. The movement of the center rod passing through the coil produces an electrical current that continuously measures the borehole diameter and is finally recorded at the surface.

The caliper log can compute hole volume for cementing purposes. Other uses include calculating annular drilling fluid velocity to determine circulation volume for lifting cuttings from the bit. They are also useful in lithologic correlations, selecting locations for setting packers, and estimating wall cake thickness.

A *temperature log* is a record of wellbore temperature versus depth. These measurements may be obtained by either electrical or self-contained temperature devices. The data from this survey shows variations in temperature. Since the curve for temperature versus depth is normally uniform, any deviation or abrupt change may indicate gas expansion or other fluid movement. This could show a casing leak or thief zone, or even a gas-bearing zone.

A *dip log* (or *dip meter log*) is a record of information dip versus depth. Both angle and direction are measured and recorded using a highly complex tool which is a variation of a microlog tool that is capable of measuring the angle of dip in the rock formations penetrated. Dip information is useful in determining hole deviation and for subsurface mapping of the rock formations as well as determining the proper direction in which to offset discovery wells and dry holes.

Other helpful logs include collar locator logs, radioactive tracers, directional logs, cement bond logs, and perforation logs. The list goes on and on.

A novice might think that much of the enormous volume of information obtained through logging is unnecessary. Keep in mind, however, that the

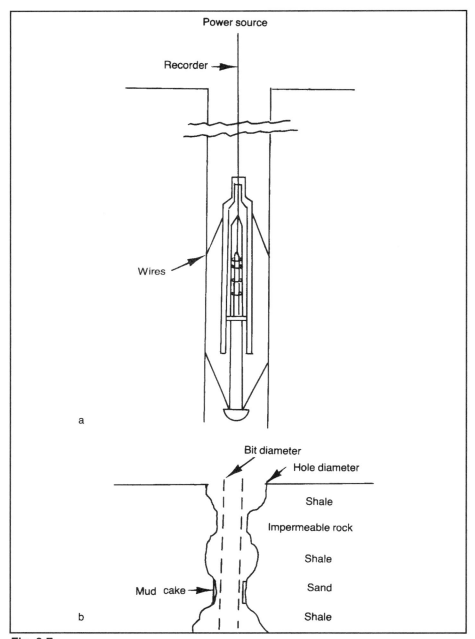

Fig. 6-7
A caliper logging tool (a) and a typical borehole, showing variations in the hole's diameter.

potential oil- or gas-producing formation lies several thousand feet below ground and in most cases cannot be examined physically. Therefore, indirect analysis methods must be used and, since substantial sums of money are at stake, it is only prudent to gather all available data that might indicate whether a well will produce.

CORING

Coring was one of the earliest formation evaluation methods. Though expensive, coring is the best method of providing samples of subsurface rock formations for detailed lithologic examination at the wellsite and for further laboratory tests.

If, on the basis of the cuttings examination, the geologist feels a formation being drilled might contain oil or gas, more positive information may be needed. One way to get such information is with cores—large pieces of the formation. For this procedure, the drillstring is removed from the hole, a core bit or sidewall corer is lowered into the hole, and the operation begins.

There are two basic types of coring operations: conventional coring and sidewall coring. Let's take a look at the two operations and note their strengths and weaknesses.

Conventional Coring

Although there are several kinds of coring tools, diamond tools are used almost exclusively for coring because of their economy. They may be used with either a core bit or with wireline coring, and their reliable cutting and recovery capability and downhole durability can reduce the required rig time to more than offset their additional cost.

A core bit is a special bit with a hole in the center (Fig. 6-8). As drilling progresses, a portion of the formation is left uncut in the center of the bit. The length of this uncut portion may range 10-80 ft, depending on the type of core barrel being used.

A wireline core barrel fits inside the drillpipe and can be removed without pulling the entire drillstring. Its cores are generally smaller, though (15-ft lengths, 1-2-in. diameter). For larger cores, a conventional core barrel is attached to the bottom of the drillstring. After the length has been drilled, the entire drillstring is removed from the hole to remove the core.

As the core sections are removed from the barrel, they are wiped, not washed, free of drilling fluid and the recovered length is measured. If the recovery does not equal the cored interval, the missing portion is assumed lost at the bottom of the hole unless there is evidence indicating otherwise. The operator, geologist, or engineer then conducts an initial examination to determine whether laboratory core analysis is warranted. If so, the core is immedi-

Fig. 6-8
A core bit and core samples.

ately placed in boxes. The boxes are marked on the end with the core number and box number. An arrow is used on the side of the box to indicate the orientation, top to bottom, of the core in the box. The depth from which the core was recovered is also shown at intervals on the side and bottom.

At the laboratory, the analysts examine the cores to determine porosity, permeability, water saturation, and oil saturation of the formation penetrated. They take this information and combine it with the wellsite geologist's data on dip, fractures, bedding irregularities, and mottling, as well as ultraviolet tests for fluorescence (under UV light, petroleum is fluorescent), for a better indication of a formation's potential production.

Sidewall Coring

Sidewall coring is a supplemental coring method used in zones where core recovery by conventional methods is small or where cores were not obtained as drilling progressed.

To secure a sidewall core, special equipment is lowered into the hole and a sample of the wall of the hole is brought back to the surface (Fig. 6-9). These samples are quite small—usually only a few inches long and about 1 in. in diameter—which makes them less reliable than conventional coring. However, a sidewall core is cheaper and takes less time. The operator must weigh the advantages and then decide which route to take.

DRILLSTEM TESTS

If the conclusions from the above kinds of tests look promising, a drillstem test can be run. A drillstem test (DST) is a temporary completion and simulates the conditions that will prevail when the well is completed.

Fig. 6-9
Schematic of a rotary sidewall coring tool (courtesy Halliburton Logging Services Inc.).

During the test, the promising interval is isolated from the rest of the bore-hole with a packer and valve assembly. Through this arrangement, fluids from the formation are directed to the inside of the drillpipe and then on up to the surface.

DSTs can be run either in cased or open holes. The principal objective is to determine the types of fluids present and the rates at which these fluids will flow.

All of these kinds of tests help the driller and engineers determine when a productive formation of commercial quantities has been penetrated. If the hole doesn't look promising, it is deemed "dry" and is plugged. On the other hand, if it shows potential, the well is completed—the subject of Chapter 7.

CHAPTER 7

Completing the Well

Formation evaluation techniques such as well logging, coring, and drill-stem testing determine whether a well can be completed for commercial production. These methods also determine certain characteristics of potentially productive rock formations to indicate the most useful method of well completion. In general, well completions are categorized as casing completions, open-hole completions, and drainhole completions. Of these three categories, the casing completion is used 90% of the time.

Casing completions can be further subdivided into the following five subcategories:

- Conventional perforated casing completions.
- Permanent well completions.
- Multiple-zone completions.
- Sand-exclusion completions.
- Water- and gas-exclusion completions.

Let's take a look at these different types of completions and see how each is accomplished and what their strengths are.

CONVENTIONAL PERFORATED CASING COMPLETIONS

This method of well completion consists of running a string of casing, or pipe, from the surface to the bottom of the hole or the bottom of the rock interval determined to be commercially productive. Then the casing is cemented in place. Often, this string of casing is named the *oil string* because petroleum is produced through it.

The oil string is secured by pumping cement down the inside of the pipe, followed by a plug that is displaced by water (Fig. 7-1). The cement is displaced to the bottom of the casing. It is then circulated up and around the outside of the casing string—the area known as the annulus. A wiper plug follows the cement. This plug fits inside the casing snugly so that the cement is wiped from

Fig. 7-1
Typical primary cementing job. In the first diagram, the cement is being pumped downhole. In the second diagram, the procedure has ended and the cement is left to harden (courtesy Halliburton).

the pipe walls as the plug is displaced by the water. The plug is stopped by a shoe, or restriction, near the bottom of the casing. The shoe keeps the cement from displacing too far up the outer annulus. When the cement hardens, the pipe is secured.

One function of the cement is to seal off any water-bearing rocks from above or below the commercially productive formation. The cement is tested for strength and sealing after it has hardened to the desired strength. The time required for the cement to harden to specifications is based on the composition of the cement mixture and the bottom-hole temperature and pressure.

To test the cement, the engineer runs a cement bond log (CBL) across the interval of cement. The top of the cement can be estimated based on the size of the hole drilled and the outside diameter (OD) of the casing. Some wells, especially shallow ones, may have cement all the way from the bottom of the casing to the surface.

A vital factor in a conventional perforated casing completion is the perforating process. Perforation—placing holes or openings in the pipe and cement—is done to establish communication (a pathway) between the wellbore and the rock formation surrounding it.

Two types of perforating guns are commonly used. The bullet perforating tool is a multibarrel firearm designed to be lowered into a well. The tool is positioned at the desired depth and is fired electrically at well from surface controls. Perforation, or penetration, of the pipe, cement, and rock formation is accomplished with high-velocity projectiles or bullets. Depending on the operator's requirements, one bullet may be selectively fired at a time or independent groups of bullets can be fired.

The other commonly used type of perforating gun is the shaped-charge perforator or jet gun, as it is often called (Fig. 7-2). In this method, the casing and cement are penetrated by a high-velocity charge of gas (approximately 30,000 ft/sec) formed by the combustion of chemical fuel inside a nozzle. This charge causes a pressure of about 4 million pounds per square inch (psi) on the target.

Two basic types of perforating tools are used: retrievable and expendable guns. A retrievable gun is composed of a cylindrical steel carrier that resembles a piece of pipe with the charges facing the perimeter of the carrier. Expendable guns are composed of materials that disintegrate into small particles when the gun is fired. The carrier is generally made of steel, but the cases housing the charge are constructed from aluminum, plastic, or ceramic. Expendable gun carriers are retrieved after firing the jet charges, which disintegrate.

Usually jet perforations are superior to bullets for penetrating dense rock formations and in multiple casing strings. Bullet penetration may equal or even exceed jet perforations in softer rocks.

Fig. 7-2
Schematic of jet perforating procedure.

Accurate depth measurements are essential to the perforating work in a well. Perforations may be placed accurately using a collar or joint locator along with radioactivity logs. The interval to be perforated is selected on the radioactivity log, and all measurements are made relative to casing collars, which are located by detectors attached to the perforating gun.

It is really better to perforate with the pressure inside the wellbore lower than the rock formation pressure. This practice lets the driller immediately remove any debris deposited in the perforation that could restrict permeability if it were to remain in place.

Swabbing

Once the casing has been perforated, the producing formation is open to the wellbore and fluids can move into the casing and up to the surface. However, the casing may be full of drilling mud. If this is the case, the well must be swabbed. Tubing is then run into the well to the depth of the producing formation.

In swabbing, a rubber plug with a check valve is lowered into the tubing on a wire line. The check valve lets fluids pass above the rubber plug as it is lowered into the hole. But when the plug is removed from the hole, the fluids cannot pass back through the valve. Thus, the rubber plug brings to the surface all of the fluids that have collected above it.

PERMANENT WELL COMPLETIONS

A permanent well completion is a technique in which the tubing is run and the wellhead is assembled only once in the life of the well. All completion and remedial operations are done using spatial small-diameter tools run inside the tubing. Perforating, swabbing, squeeze cementing (sealing off leaks in the casing), gravel packing (using gravel to fill a wellbore to prevent cave-ins or encroaching sand), and other completion and remedial work can be done through the tubing. The advantage of this technique is economy.

Let's take squeeze cementing for an example. In squeeze cementing, a tubing extension is run on a wire line and is packed off inside the tubing near the bottom of the tubing. After the depleted zone is squeeze cemented, the excess cement is circulated out of the hole. The tubing extension is removed, and other operations may be continued, such as perforating a new zone uphole using a through-tubing perforating tool.

In a conventional recompletion, mud would have to be pumped into the well until the pressure died, the tubing would have to be pulled and rerun with a cement retainer tool, the tubing would have to be pulled again and a perforating tool run, the casing would be perforated, and finally the tubing could be rerun one last time to produce the hole. With a permanent well completion, you don't have this expensive work. However, the tools used in permanent well completions are small and less efficient, and they are more prone to failure than the normal-sized tools used on conventional completions.

MULTIPLE-ZONE COMPLETIONS

In some areas, more than one interval may be productive in a single well. A multiple-zone completion allows simultaneous production of two or more separate productive intervals. This is often demanded of operators by regulatory agencies who want to keep various classifications of oil separate. It can also be required for reservoir control—a high-pressure formation and a low-pressure formation.

Fig. 7-3
Dual completion well. With this method, two producing formations can be tapped with only one wellbore.

Two-zone, or dual, completions are the most common type (Fig. 7-3). Triple- and quadruple-zone completions have also been performed, but less commonly. It's easy to see one disadvantage to the technique: the more completions, the more complex and (expensive) the downhole equipment and tools needed to achieve and maintain zonal segregation. The problems are enhanced when one or more intervals require artificial lift (pumping, etc.) in order to produce.

The oil industry has been reluctant to accept multiple-zone completions. The interval savings enjoyed by eliminating the need for a separate well for each productive interval generally has been offset by production and remedial problems encountered after completion. The initial economy does not always provide a profit.

Overall, the technique depends entirely on the total economics compared with the economics of separate wells for each interval. Certain lease obligations may be fulfilled by using multiple-zone completion techniques at less expense than a separate well for each interval. Field development may be accelerated using this technique. During times of materials shortages, costs of steel pipe may rise sharply. Continued improvement in technology may enhance the use of this completion technique.

SAND-EXCLUSION COMPLETIONS

When a well is dug in unconsolidated (loose-grained) sand, the completion is much more complicated than the two previous types of completions. Sand production can erode the equipment and wellbore and plug the flow lines so much that well operation becomes uneconomical. At very low rates, little or no sand may be produced; at high rates, however, large quantities may be carried along in the production stream.

Early in the industry, sand production was tolerated in flowing wells. The only control measure was some means to prevent accumulation. When it became necessary to pump the wells, exclusion methods were required to prevent pumping equipment erosion. Today, many wells now produce that would be uneconomical without sand control.

Two completion techniques used to control sand are using slotted or screen liners and packing the hole with aggregate such as gravel (Fig. 7-4). The principle behind these two methods is that the openings through which the fluids must flow must be the proper size. When this happens, the sand forms a bridge and is excluded.

The first step is to obtain a sample of the formation sand and analyze it for grain size. This helps decide the size of the slots or screens and the size of the aggregate or sand grains. After the slotted or screened liner is constructed, based on the screen analysis, the liner generally is run into the well on tubing (small-diameter pipe) and hung with a liner hanging tool opposite the productive interval. This can be done in either a cased hole or an open hole.

Gravel packing may be performed in several ways in either perforated or open-hole intervals as well. Here, the liner slots or screen is used only to exclude the gravel. The slots may be larger than for the previous method and are generally only slightly smaller than the aggregate or gravel. The thickness of the gravel pack is usually four or five gravel diameters. As mentioned earlier, the formation sand bridges with the pores of the gravel pack, and the gravel is prevented from entering the screen by the liner.

The sand-exclusion completion technique can be applied at the time of completion or later in the life of the well when an unexpected problem occurs. In many parts of the world, sanding is not a major problem. In California and on the Gulf Coast, however, sand problems occur almost daily and are handled routinely.

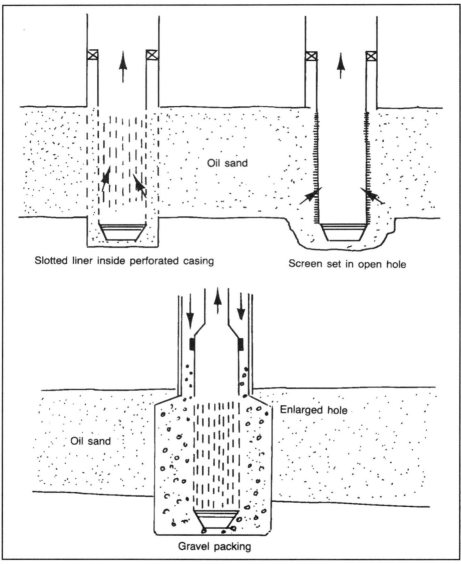

Fig. 7-4
Two sand control techniques: (a) liners and (b) gravel packing.

WATER AND GAS-EXCLUSION COMPLETIONS

Generally, operators want to produce as little as possible with oil. Water must be separated from oil before the crude can be sold, and the more that must be separated, the less oil there is to be sold. Reducing or excluding the amount of gas produced is also desirable, unless the well is completed in a gas-only reservoir. Gas is the reservoir's energy, the force that pushes the fluids into the wellbore. Therefore, that energy source should be conserved for as long as possible to enhance the life of the field.

In many reservoirs, the oil zone occurs with an overlying gas zone or an underlying water zone, or both. In these situations, the producing wells must be completed so free gas and water are not produced. Selecting the proper interval for production within the zone of interest is thus important to limit gas and water production.

Let's digress for a moment and study how reservoir fluids behave during production. In order to produce the well, a pressure *sink* or gradient must be established between the well and its drainage radius, the area around the well that contains hydrocarbons. This pressure gradient extends horizontally and vertically. Thus, the pressure sink created by the well acts on all three fluids in the reservoir: the oil, the gas, and the water. Consequently, all three fluids then to flow toward the well. Water is denser than oil, and oil is denser than gas. These density differences furnish an opposing gradient that tends to prevent the water from rising above its static level. If the producing rate is not too high, the oil-water interface (the division between the oil and water intervals) will merely rise until it reaches an equilibrium position such that the two opposing gradients are equal. The reverse occurs between gas and oil; the gas-oil interface drops until the two opposing gradients are equal.

However, if the producing rate becomes too great, the water and the gas may both be drawn into the well. This results in water production and its attendant problems, including the expense of disposal and/or free gas production, waste of reservoir energy, and consequent rapid decline in reservoir pressure. Further deterrent are the penalties that regulatory agencies impose for producing excessive gas. So the result is that producers want to exclude water and gas production whenever possible.

One of the more useful survey tools for determining water entry into a wellbore is a radioactive tracer survey tool. This tool is run to the bottom of the well while the well is on production. A water-soluble radioactive tracer is released. As the tracer moves uphole, it mixes with the water as the water enters the wellbore. The tool picks up the radioactivity of the tracer and follows it as it moves uphole. The location of the point of entry can be observed at the surface and recorded on a graph inside the survey service company van.

In reservoirs composed of alternating layers of productive sands separated by shale or dense sections, squeeze cementing and perforating often reduce or exclude water production. Sometimes the water saturations in the reservoir rock are so high that water exclusion is impossible, and water must be produced along with the oil.

Open-Hole Completions

Open-hole, or barefoot, completions are situations in which the oil string is set on top the indicated productive interval, leaving the productive interval as an open borehole with no pipe to protect it (Fig. 7-5). This method can only be used in highly competent rock formations that are not subject to collapse.

Often this completion method is used in low-pressure, hard-rock areas where cable tools are used for the drilling-in operation. Rotary drilling tools are used until the oil string is set. At that time, the rotary tools are moved out and the cable tool rig is moved in. The cable tool rig then bails out the mud and drills the desired productive interval.

This method is good because the interval can be tested while it is drilled. The tools do not have to be pulled, casing cemented, and well perforated. It also eliminates formation damage caused by drilling mud and cement and allows incremental deepening as necessary to avoid drilling into water. This last point is important in thin water-drive reservoirs where the productive interval is only a few feet thick.

Often a downhole stimulation method may be used to speed up the rate of flow from the productive interval. The most common stimulation techniques include shooting with nitroglycerine (now outdated), hydraulic fracturing, and acid fracing. More about these methods in Chapter 9.

It's easy to see that an open-hole completion is more productive than a conventional perforated-casing completion in which the fluids must enter the wellbore through a few small-diameter holes in the pipe. It is especially advantageous in thin, laminated rock strata or other situations where vertical permeability is either low or discontinues.

The open-hole method is also less expensive since some casing is eliminated and perforating costs are excluded. Contamination by cement can be avoided, as well as damage to drilling fluids. However, the perforated-casing completion offers a much higher degree of control over the productive interval, since an interval can be perforated and tested as desired. Individual sections or intervals can be isolated and selectively tested and stimulated more easily and satisfactorily through perforations than in an open hole. Also, hydraulic fracturing is more successful in perforated casing completions. Productivity is generally about 50% higher than for open-hole completions. Improved zonal control is also of substantial value when remedial measures, such as water or gas exclusion, are undertaken.

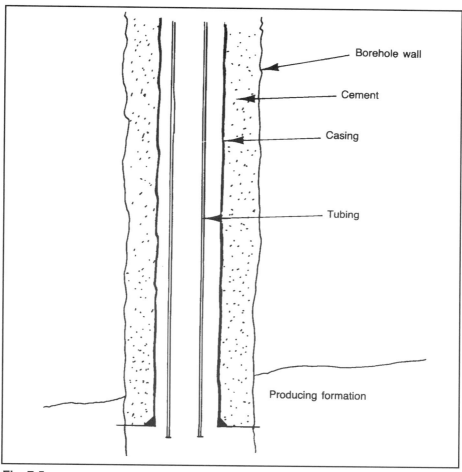

Fig. 7-5
An open-hole or barefoot completion.

DRAINHOLE COMPLETIONS

Drainhole completions is a broad term applied to a number of well completion techniques. In general, the term applies to a well that has been drilled and completed in some form of horizontal or near-horizontal situation. This technique generally requires the use of some form of directional drilling—drilling at an angle rather than straight down (Fig. 7-6).

The basic principle involved in drainhole completions is to deviate the hole vertically as the well is drilled, increasing the deviation until the hole is

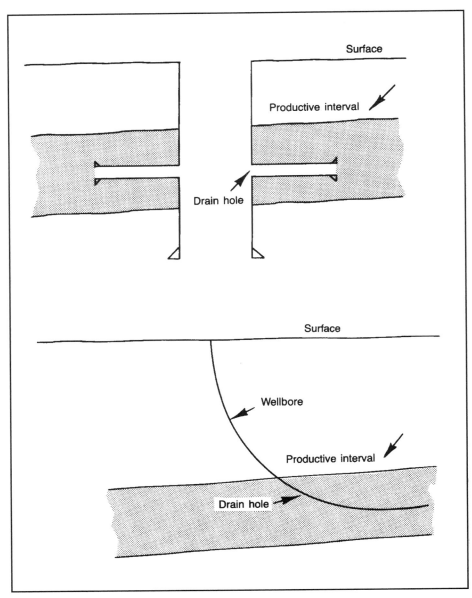

Fig. 7-6
Two types of drainhole completions. In the first, drilling is horizontal into the produc-ing formation from a vertical borehole. In the second, a hole is drilled at an angle to penetrate the producing formation horizontally.

Fig. 7-7
Comparison of types of sand exclusion completions.

nearly horizontal as it enters the productive interval. The result is a long productive interval in the well, which results in increased productivity.

Other types of drainholes are accomplished by drilling one or more lateral offshoots from the main borehole. These offshoots, or auxiliary wellbores, are also referred to as drainholes. Some of the main wells are as large as 8 ft in diameter. These auxiliary wellbores are drilled by people entering the well and working in the bottom of the well, drilling the offshoots like miners.

The balance between the additional cost of drilling and completing the drainholes should be analyzed, considering the additional productivity achieved. With continued improvement in technology and cost reduction, applications may increase. However, drainholes must compete with normal stimulation techniques capable of obtaining similar productivity increases in most areas.

Throughout this chapter, we've talked about the difference between cased-hole and open-hole completions. Let's take a closer look at casing and cementing procedures before we continue with different kinds of production methods.

CHAPTER 8

Casing and Cementing

Very often in completions, casing must be run to seal the wellbore from encroaching fluids. In order to attach the casing firmly to the wellbore wall and stabilize the hole, cement is pumped downhole. We touched on these methods in Chapter 7. Let's take a closer look now at these important techniques.

CASING

Casing must be run into a well if commercial indications of hydrocarbons are observed. Casing normally is run through the lowest interval that has potential; then it is cemented in place.

Casing has several functions:

- It contains formation pressures and prevents fracturing of the upper and weaker zone.
- It keeps the hole from caving in.
- It confines production to the wellbore.
- It provides an anchor for surface equipment.
- It provides an anchor for artificial lift equipment.
- It separates the formations behind the pipe and limits production to the zones selected by the engineer.

Because casing has several different functions, it is usually necessary to install more than one string of casing or pipe. These kinds of casing are divided into five classifications:

- Conductor pipe.
- Surface casing.
- Intermediate casing.
- Liner string.
- Production casing.

Conductor Pipe

In certain areas, because of the nature of the surface soil, it may be necessary to install a short length of conductor casing, usually not more than 20-50

101

ft long, at the surface to prevent excessive caving around the sides of the hole. Conductor pipe is the conduit that also raises drilling fluid high enough above ground level to return the fluid to the mud pit. And it prevents washing out around the rig's base.

Conductor pipe is set after the well location has been graded and prepared for the rig. The mud pits, if any, are dug. The hole for the conductor pipe is drilled with an auger drill mounted on the end of a truck. Then the pipe is lowered into the hole and concrete is poured around it to fill the surrounding space.

In swamps and offshore locations, the pipe is driven in with a pile driver. Offshore, the diameter of the pipe can range 30-42 in., while onshore diameters are usually smaller—16-20 in.

Surface Casing

The next casing to be set is surface casing, which protects freshwater formations from contamination by oil, gas, or salt water from the deeper producing formations. Since freshwater formations normally occur at shallow depths, no more than 2,000 ft of surface casing are usually required.

An important auxiliary function of the surface casing is to provide a place to attach the blowout preventers (BOP). Blowout equipment is attached during drilling operations to contain kicks or pressure surges downhole. Once the well is completed, a production manifold, or *christmas tree* (Fig. 8-2), replaces the BOP.

The surface casing should be set deep enough to reach rock formations that will not fracture or break down with the maximum expected mud weight at the depth where the next string is to be set. The outside diameter of the surface string is slightly smaller than the inside diameter of the conductor pipe. (The surface casing is lowered inside the conductor pipe.) The minimum depth is usually 10% of the expected total depth (TD) of the well or 500 ft, whichever is greater. When the expected depth is reached, this string of casing is cemented to the surrounding conductor pipe, which anchors it in place.

Intermediate Casing

An intermediate casing, though not always run, protects the hole against loss of circulation in shallow formations. When drilling in areas that have abnormal formation pressures, heaving shales, or lost-circulation zones, a string of casing may need to be run to minimize hazards before drilling to greater depths. That's the purpose of intermediate casing. Strictly speaking, it is not required for the well to function properly, and it is more a part of the drilling operation than the completion operation.

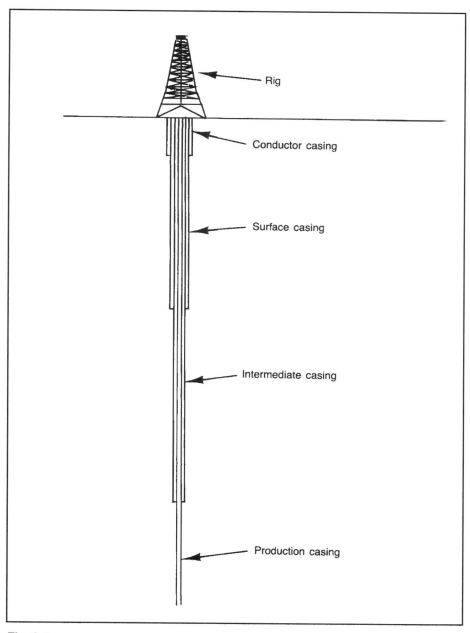

Fig. 8-1
Four classes of casing.

Fig. 8-2
A christmas tree.

Intermediate casing strings are suspended and sealed at the surface with a casing hanger. The lower portion is cemented by circulating cement down and out around the bottom of the pipe and up across the intervals where cement is needed. More about cementing will come later in this chapter.

Liner Strings

Unlike casing that is run from the surface to a given depth and overlaps the previous casing, a liner is run only from the bottom of the previous string to the bottom of the open hole. Liners are suspended from a previous string with a hanger. They are often cemented in place but may be suspended in the well without cementing.

One advantage of using a liner is that it is not necessary to run the string back to the surface. Casing is expensive, and substantial savings may be realized by using less.

Sometimes liners are set in a hole as a protective string, serving the same function as an intermediate string.

Production Casing

Production casing is sometimes known as the oil string or the long string. It isolates the oil and/or gas from undesirable fluids in the producing formation and from other zones penetrated by the wellbore. This casing also serves as the protective housing for the tubing and other equipment used in a well.

The oil string is the last string of casing run in the well. It is a continuous length of pipe from the well surface to the producing formations.

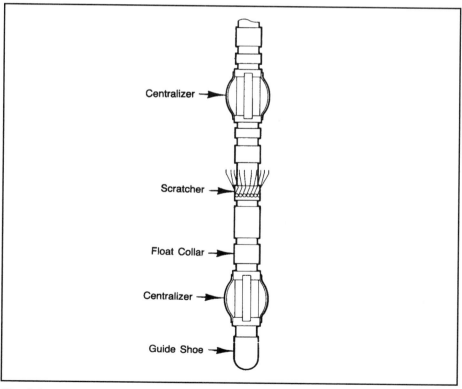

Fig. 8-3
Well completion equipment used when casing is run into the hole (courtesy Trico Industries).

Fig. 8-4
Close-ups of well completion equipment: (a) guide shoe, (b) float collar, (c) centralizer, (d) scratcher (courtesy Halliburton).

CASING ACCESSORIES

Many kinds of tools and accessories are used when casing is run into the hole (Fig. 8-3). Let's take a look at a few of these.

Guide Shoes

The guide shoe is a heavy, blunt object placed on the bottom of the casing which prevents the lower end of the pipe from deforming. A guide shoe has a rounded nose to guide the casing through and around any obstacles encountered in the well. It also has a cement collar inside the bottom to aid in bonding cement to the casing string. The interior diameter (ID) is smaller than the casing to restrict drilling fluid from rising into the casing and to provide some amount of flotation.

Float Collars

Float collars are multipurpose devices that permit the casing to float into the wellbore. They are equipped with a back-pressure valve that is closed by the pressure outside the fluid column. This valve prevents fluid entry as the casing is lowered into the hole. The valve also serves as a check valve in the string to prevent cement from back-flowing after being pumped up the annulus and outside the string of pipe. This device is important because the density of the slurry is always greater than the drilling fluid's density.

The back-pressure valve also prevents a blowout through the casing if a kick should occur during the cementing operation. This safety feature is extremely important when high-pressure formations are exposed in the open hole.

A float collar can serve as a stop for the top cement plug when cement is displaced. The advantage is that a known quantity of cement slurry will remain inside the casing between the float collar and the casing shoe. If the circulation of cement slurry is topped inside the casing, a good-quality cement will fill the annular space above the casing show. This lets some of the cement slurry remain inside the string of pipe at the casing shoe and gives the operator reasonable assurance that good-quality cement is outside the casing at that point.

Centralizers and Scratchers

Centralizers and scratchers are attached to the casing to aid the cementing process. Centralizers are still springs that hold the pipe in the center of the hole away from the wellbore. This ensures adequate distribution of cement around the pipe. Scratchers are mechanical devices with wire fingers attached to the pipe along with the centralizers. The scratchers abrade the hole by reciprocating or rotating the casing string. Mud cake—a coating of drilling fluid and drilling particles that build up on the wellbore wall—can then be removed from the hole. This provides a better surface to which the cement can bond.

Wellhead

The wellhead is the casing attachment to the BOP or the production christmas tree. It is a permanent fixture that is bolted or welded to the conductor pipe or surface casing. The wellhead is located in the cellars of land wells and in the cellar deck of jack up or other offshore platform rigs. Barges, semisubmersibles, and drillships install the wellhead on the sea bed.

The surface casing nearly always is welded to the wellhead. Subsequent casing strings are inserted inside a wellhead housing and are supported in a casing hanger, which latches and seals inside the wellhead housing. Wear bushings or bore protectors protect the sealing surface while drilling through the wellhead.

CEMENTING

It is easy to see that the wellhead could not support the weight of thousands of pounds of casing. At a certain point, the casing must be attached to the wall of the wellbore for additional stability. That's one role of cementing.

Oil well cementing is the process of mixing and displacing cement slurry down the casing and up the annular space behind the pipe. A bond between the pipe and the formation is made after the cement sets.

Cementing serves several purposes:

- Bonds the pipe to the rock formations.
- Protects the pipe and the producing formations.
- Seals off troublesome formations before drilling deeper.
- Helps keep high-pressure zones from blowing out.
- Provides support for the casing.
- Prevents pipe corrosion.
- Forms a seal in the event of a kick (sudden pressure increase) during further drilling.

Cementing is classified as primary or secondary. Primary cementing is done immediately after the casing is run into the well. The objective is to effectively seal and separate each zone and to protect the pipe. Secondary cementing is performed after the primary cement job. Usually it is part of a repair or remedial job.

Primary Cementing

There are seven primary cementing methods:

- Single-stage cementing, through casing, called a normal displacement technique.
- Multistage cementing, used in wells that have critical fracture gradients or that require good cement jobs on long casing strings.
- Inner-string cementing through drillpipe for large-diameter pipe.
- Multiple-string cementing for small-diameter pipe.
- Reverse circulation for critical formations.
- Delayed setting for critical formations and to improve placement.
- Outside or annulus cementing through the tubing for surface pipe and other large-diameter pipe.

Of these seven methods, single-stage and multistage cementing are the primary ones.

Single-Stage Cementing

In practice, a spacer of 10-15 bbl of water or chemical is pumped behind the drilling fluid and ahead of the bottom plug. The water or chemical serves

Fig. 8-5
Overview of a cementing job.

as a flushing agent and provides a space between the mud and the cement slurry. It also helps remove the mud cake from the wellbore and flushes the mud ahead of the cement slurry, thereby reducing contamination.

Cementing plugs usually are an aluminum body encased on molded rubber. When the bottom plug reaches the float collar, the diaphragm ruptures to permit the cement slurry to proceed down the casing and up the annular space outside the pipe. The top plug, which is solidly constructed, is released when all the cement has been mixed. This plug follows the cement slurry. The cement slurry is chased by drilling fluid or other fluid to displace the cement down the casing. The plug causes a complete shutoff when it reaches the float collar. A plug containing a cementing head is used to release the plugs.

An increase in pump pressure is a signal that the top plug has reached the float collar. This is called *bumping the plug*. To ensure good cement circulation and displacement, the casing should be moved either by reciprocation or rotation, or both, and this should be continued throughout the time needed for circulation, cement mixing, and displacement.

Multistage Cementing

These techniques are used for cementing two or more separate sections behind a casing string. This is usually for a long column that might cause formation breakdown if the cement were displaced from the bottom. The essential tool consists of a ported coupling placed at the proper point in the string.

The lower portion of the casing is cemented first in the usual manner, using plugs that will pass through the stage collar without opening the ports. The multistage tool is then opened hydraulically by special plugs. Then fluid is circulated through the tool to the surface. Cement is placed in the upper section through ports that are subsequently closed by the final plug behind the cement.

Secondary Cementing

Secondary cement work is done after a primary cement job. It may be used to plug an open zone, to plug a dry hole, or to squeeze cement through perforations. Squeeze cementing segregates oil and gas producing zones from formations containing other fluids. It is also used to:

- Supplement or repair a primary cement job.
- Repair defective casing or improperly placed perforations.
- Reduce the danger of lost circulation in an open hole while drilling deeper.
- Abandon a permanently nonproductive or depleted zone.
- Isolate a zone before perforating.
- Fracture the formation.

The slurry is injected under pressure through the perforations. The pumping rate is slow to allow for dehydration and initial setting. Pumping continues until the desired squeeze pressure is reached.

CEMENT CLASSIFICATIONS

As a result of effort to find hydraulic cements that could be used underwater, scientists discovered that the limes produced from impure limestones would yield mortars superior to those produced from pure limestones. This

Fig. 8-6
How cement is run through the casing and annulus (courtesy Halliburton).

discovery led to the burning blends of calcareous and argillaceous materials—a patented process that yielded a material known as Portland cement, which resembles concrete produced from the Isle of Portland off the coast of England.

The Portland cements used for oil well cementing carry API classifications (Table 8-1). The usable depth of API cements is a function of temperature and pressure. In areas of sub-normal temperatures, API cements can be used at greater depths. In abnormally high temperatures, they may be limited to shallow depths. Normal API temperature gradient is considered to be 1.5∞ F/100 ft of depth.

CEMENT ADDITIVES

Most cementing jobs are performed using bulk systems rather than handling sacks manually. Bulk systems let workers prepare and supply compositions tailored to suit the requirements of any well condition. This is accomplished by using additives with API classes A, B, G, or H cements. Some of the additives are retarders; some are accelerators that alter cement setting times. Various additives can provide the following functions:

- Reduce slurry density.
- Increase slurry volume.
- Increase thickening time and related setting.
- Reduce waiting-on-cement (WOC) time and increase early strength.
- Reduce water loss.
- Help prevent premature dehydration.
- Increase slurry density to restrain pressure.

Conductor and surface casing cements have lower temperatures and require an *accelerator* to promote setting of the cement and to reduce excessive waiting time.

For deep wells, cement *retarders* help extend the cement's pumpability. The primary factor that governs the use of additional retarders is the well's bottomhole temperature. As temperature increases, the chemical reaction between cement and water is accelerated. This reduces the cement's thickening time and pumpability. Pressure has some effect, but and increase in temperature of as little as 20∞F may mean the difference between an unsuccessful and a successful cementing job.

Lightweight additives reduce slurry density. *Heavyweight additives* are used when abnormally high pressures are expected.

Lost circulation is a common drilling problem. The same problem can occur during cementing operations, and it may be necessary to use cements containing *lost-circulation additives* to maintain circulation.

Mixing low-water-loss additives into oil well cements to reduce filtration rates is similar to the procedures used in drilling fluids. *Fluid-loss additives* are sometimes used in squeeze cementing and in high-column cementing such as deep liner cementing.

Contrary to the practice in pumping drilling fluid, cement should be in turbulent flow to ensure better flushing of mud from the annulus. Lower-viscosity slurries will go into turbulent flow at lower pumping rates. This reduces circulation rates and allows cement to be pumped in turbulent flow at less than formation breakdown pressure. *Friction-reducing additives* promote turbulent flow at low displacement rates.

Salt-saturated cements were developed for cementing through salt zones because fresh water does not bond properly to salt formations. The water from the cement slurry dissolves or leaks away the salt at the interface, which prevents an effective bond. Salt slurries also help protect shale sections sensitive to fresh water.

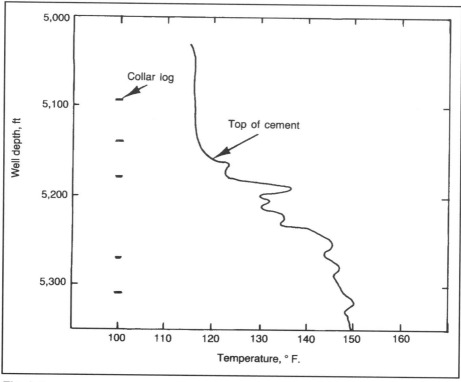

Fig. 8-7
Temperature survey showing Top of Cement.

CEMENTING CONSIDERATIONS

Oil well cement slurries range in density from 10.8 to 22 lb/gal. The slurry density depends on the amount of mixing water and additives in the cement and the amount of slurry contamination from drilling mud or other foreign material. Slurry density control is usually maintained by measuring the density of the cement passing through the mixing tub with a standard mud balance.

The volume of cement required for a cementing job is based on calculated volumes, field experience, and regulatory requirements. In the absence of experience in an area, a volume of 1.5 times that calculated from the wireline caliper survey or estimated from bit measurements is used.

As cement sets, its temperature increases considerably. This phenomenon can be used to determine exactly where the top of the cement is located (Fig. 8-7). Once the cement has begun to set, a recording thermometer is lowered into the hole and a temperature log is taken. From the top of the cement to the bottom of the hole, the temperature will be substantially higher than it is above the cement.

This log can also determine the quality of the bond between the casing and the wellbore. A poor bond is shown as a variation in temperature that is out of line with the normal temperature gradient. A more sophisticated tool is the *cement bond log* (CBL), which measures the diminished strength of acoustic signals. The CBL can determine the cement bonding to both the casing and the rock formation. It requires skilled interpretation and, under favorable conditions, can even determine the cement's compressive strength.

Once the casing has been set, cemented, and perforated and any necessary stimulation has been performed, the well is ready to be equipped for production. The next chapter deals with getting the hydrocarbons from the producing zone to the point of sale.

CHAPTER 9

Production Concepts

Efficient oil recovery from a reservoir demands a basic working knowledge of fluid mechanics and specific recovery processes. Then this knowledge must be applied to each reservoir. For this, an engineer must recognize the individual characteristics of the reservoir, the process best suited for a chosen reservoir, and the operation methods that ensure the reservoir yields the maximum possible amount of oil.

OIL RECOVERY MECHANISMS

Oil recovery is a displacement process. Oil does not have any inherent ability to expel itself from a reservoir. Rather, it must be displaced from a porous formation to a wellbore via some displacing agent. Generally, gas or water is used as this agent, and often one or both of these are available within or near the reservoir. If they aren't, the operator can supply the gas or water through injection wells.

We've already discussed that the three major natural oil displacement mechanisms are dissolved-gas drive, gas-cap drive, and water drive. The type of drive chosen fixes the operating characteristics and to a large extent determines the ultimate recovery. All three methods differ in their characteristics, mechanisms, and efficiencies.

In a *dissolved-gas drive,* oil is displaced when the gas is liberated from solution in the oil. This happens when production reduces the pressure in the formation. Present understanding indicates that this drive is basically inefficient because depletion occurs simultaneously throughout the reservoir.

In a *gas-cap drive,* the agent is the free gas cap originally present and overlying the oil-bearing zone. In this mechanism, reduced pressure results in the gas cap expanding. As the gas cap expands downward and invades the oil zone downstructure, it drives the oil toward the regions of reduced pressure—the producing wells.

In a *water drive*, water from adjacent aquifiers encroaches into the oil-bearing portion of the reservoir. In response to the reduced pressure from the wellbore, the water flows in the direction of the reduced pressure, invades the oil zone, displaces the oil from the porous rock, and drives the oil ahead of it toward the well.

Gas-cap and water drives are much more efficient than dissolved-gas drives, with water drive generally the most efficient. But often nature must be supplemented or even modified to ensure maximum efficient recovery.

In each of these recovery mechanisms, gravity exerts a supplemental effect upon the process and must be considered when vertical direction is involved. Under some conditions, gravity can be a primary agent for oil recovery. In cases where fluid movements are at rest or only slightly disturb pressure, the combined effects of gravity and pressure can cause the different fluids to segregate by relative densities. Since oil is lighter than water, oil may move ahead of the displacing water and increase production.

The conditions where gravity may be important to recovery include permeable, thick formations with a reasonable degree of tilt, low oil viscosity, and a rate of production low enough to diminish any disturbance.

FACTORS INFLUENCING RECOVERY

The amount of oil that can be recovered from a reservoir varies widely, depending partly on the natural conditions imposed on the underground structure and partly on the properties of the fluids. These are further subject to the operator and how he develops the field. Among those factors that may exert an influence on oil recovery are the following:

- Characteristics of the productive formation, e.g., porosity, permeability, interstitial or connate water content, and uniformity, continuity, and structural configuration.
- Properties of the reservoir oil, i.e., viscosity, shrinkage, quantity of gas in solution.
- Operation controls, i.e., control of natural expulsive forces, rate of production, pressure behavior.
- Well conditions and structural location.

CONTROLLING RESERVOIR PERFORMANCE

Each kind of drive requires careful control to avoid wasting the displacing agent. The operator can exert some control over nature's forces, especially in direction. The problem is whether the operator can substitute a more efficient drive for a less efficient one and then control it for maximum return.

The important consideration is early planning. After a new reservoir is discovered, the operator should determine the type of drive naturally available and its efficiency. With this information, the operator can decide whether to take full advantage of the natural drive, to supplement the natural drive, or to modify things completely by injecting gas or water. This decision is balanced by considerations for total development and operating costs. In the development aspects, the arrangement, location, and manner of completion is much different for a gas-cap well than for a water drive. But under either drive it is usually better to operate at a high level of reservoir pressure. This means starting the operation as early in the life of the pool as possible.

Requirements for Proper Control

Efficient recovery depends on the degree to which the advancing gas or water invades the entire reservoir and how uniformly the gas or water displaces or flushes the oil. There are seven basic requirements for controlling an oil reservoir properly:

- Choose an efficient dominant mechanism for recovery. It may be only the natural drive, it may be supplemented with injected fluid, or it may be modified to create a completely new drive.
- The dominant mechanism must consistently and progressively advance the fluids throughout the entire reservoir, with the invading fluid displacing the oil ahead of it to the producing wells.
- The boundary between the invaded and uninvaded portions of the reservoir must be sharply defined and at all times reasonably uniform.
- Flush the oil uniformly. Highly oil-saturated zones must not be trapped or bypassed behind the advancing gas or water front.
- Avoid excessively dissipating the gas or water.
- Locate and complete wells for adequate control of the advancing gas or water.
- Maintain reservoir pressure high enough to prevent excessive release of solution gas.

CONTROLLING RATE OF PRODUCTION

Efficient recovery doesn't come by chance; it takes careful and deliberate action on the part of the operator. One of the most essential factors for efficient recovery is controlling the rate of production. Studies show that excessive production rates result in rapid declines in reservoir pressure, premature release of dissolved gas, irregular movement of the displacement fronts, dissipation of gas and water, trapping and bypassing oil, and in extreme cases an inefficient dissolved-gas drive. Each of these, resulting from excessive withdrawal rates, reduces the ultimate recovery of oil. Generally, the most effective way to control recovery is to restrict the production rate.

Rate control alone, though, may not be enough. The operator may also have to control the progress of the displacing fluid and prevent its premature dissipation. So conservation measures must be taken to prevent waste and control reservoir performance.

MAXIMUM EFFICIENT RATE

Recovery from most pools is directly dependent on the rate of production. For each reservoir, there is a maximum rate of production that will fulfill the efficient recovery. Increasing production beyond this maximum usually leads to loss of drive and reduces ultimate production. On the other hand, producing below this maximum rate will not increase the ultimate oil recovery. From these considerations has developed the concept of the maximum efficient rate of production (MER).

The maximum efficient rate for an oil reservoir is defined as the highest rate that can be sustained for an appreciable length of time without damaging the reservoir and which, if exceeded, would minimize ultimate oil recovery. This concept has a sound basis as an engineering principle. MER is not an invariable characteristic of a reservoir but depends on the recovery mechanism and the physical nature of the reservoir, its surroundings, and its contained fluids. For the same reservoir, the MER differs from one recovery process to another. But through studying the reservoir and behavior, engineers can determine the MER if enough geologic and operating information is available.

In establishing the maximum efficient rate for a reservoir, two independent physical conditions and one economic condition must be satisfied:

- The rate must not exceed the capabilities of the reservoir.
- The individual well rate must not be excessive.
- The individual well rate must not be so low as to prohibit profitable operation.

In the early stages of field development, MER is usually limited by the efficient rate for the individual wells. After development is essentially complete, there are usually enough wells to produce the total reservoir MER without simultaneously exceeding the individual wells' MERs. Hence, in the later stages of development, MER is limited by the reservoir's efficient capacity. In any case, the small of the two capacities— either of the reservoir or of the individual wells—fixes the MER for the field.

EFFICIENT WELL PERFORMANCE

Efficient reservoir production also demands efficient operation of the wells tapping the reservoir. We just learned that the MER for a reservoir cannot exceed the combined MERs of the individual wells. Thus, to determine the efficient capacity of a reservoir means the operator must investigate the capabilities and limitations of each well to produce its share.

One of the most useful tools in determining productive capacity of a well is the flow test. The flow test determines the productivity factor and the specific productivity factor of the well. These data in turn give the total pressure drop and the pressure drop per unit of formation section open to a well during flow at a given production rate. So this test evaluates the maximum rate at which a well may be produced to avoid excessive localized pressure drops around the well, to maintain high oil saturation, and to prevent gas or water fingering, or encroaching, into the well.

Well potential tests, production tests at regular intervals, and continuous records of well production histories also give information valuable in assigning efficient producing rates to individual wells.

Now we know how reservoirs work and why maximum efficient recovery rate is important in obtaining the ultimate producibility from a well. When these factors are determined, the operator decides which drive to use for the well and/or reservoir. At this point, the operator is ready to equip the well with production equipment and can begin producing oil.

chapter 10

Production Methods— Equipment and Testing

Once the casing has been set, cemented, and perforated and any necessary stimulation has been performed, the well is ready to be equipped for production. This chapter deals with getting the hydrocarbons from the producing zone to the point of sale.

EQUIPPING THE WELL FOR PRODUCTION

To protect the casing, a smaller-diameter string of steel pipe called *tubing* is run into the well. Through this tubing the well fluids will be brought to the surface. To keep the well fluids out of the annular space between the casing and tubing, a *packer* that can be expanded to form a leakproof seal is usually placed near the bottom of the tubing string.

A number of valves and fittings must be installed at the top of the well to control and direct the flow of well fluids.. As previously mentioned, this assembly of valves and fittings, called the wellhead, is sometimes referred to as a christmas tree.

From the wellhead, the produced fluids are transported through a flow line to a field gathering station, usually called a *tank battery*. In these tanks the production from many wells may be collected. The tank battery also contains the equipment necessary to separate the produced fluids—oil, water, and gas— so each can be handled appropriately. More about this will be detailed in Chapter 11. First, let's discuss the procedures and equipment required to get the production from the bottom of the well into the flow line.

CLASSIFICATION OF WELLS BY TYPE OF LIFT

Producing wells are normally classified by the type of mechanism used to get the produced fluids from the bottom of the well into the flow line. This mechanism may be natural flow or some type of artificial lift. Gas wells are produced by natural flow. Some oil wells flow naturally due to inherent energy during the early stages of their productive life (Fig. 10-1), but eventually they require energy from an external source to maintain productivity.

When a well is opened to production, the oil enters the wellbore by virtue of the difference in pressure between the wellbore and the reservoir. As the pressure is reduced, the solution gas begins to vaporize, forming bubbles within the oil. As the oil flows up the tubing string, the pressure further diminishes. These gas bubbles expand, lightening the fluid column. The combination of reservoir pressure and the reduced weight of the fluid column in the tubing string permits the well to flow naturally.

During the course of production, gas bubbles are also forming within the reservoir itself. The bubbles continue to expand, forcing more oil into the

Fig. 10-1
Schematic drawing of a flowing well with sufficient pressure to lift the oil to the surface.

wellbore. Eventually, however, the expanding gas bubbles interconnect, forming continuous channels of gas through the reservoir. As this happens, gas begins to flow into the wellbore, leaving much of the heavier oil behind (Fig. 10-2). These phenomena continue until the reservoir pressure is too low to force the remaining, heavier oil to the surface. At this point, artificial lift is required.

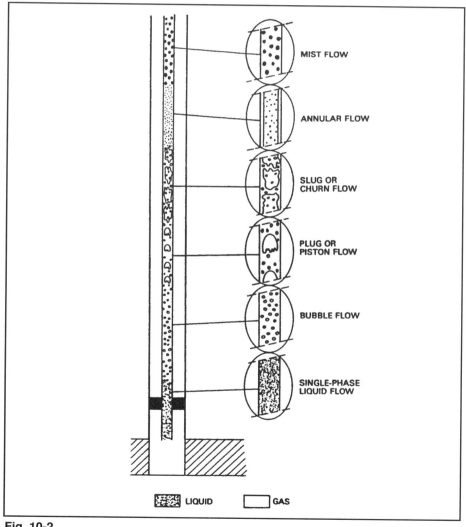

Fig. 10-2
Separation of fluids in the tubing while flowing.

ARTIFICIAL LIFT

Artificial lift is used when the pressures in the oil reservoir have failed to the point where a well will not produce at its most economical rate by natural energy. The most common methods of artificial lift are:

- Gas lift.
- Plunger lift.
- Sucker rod pumping.
- Pneumatic and hydraulic pumping.
- Rotary positive displacement pumping.
- Hydraulic pumping, subsurface.
- Electric submersible pumping.

In achieving the maximum economic gain in oil production, one must consider the time value of money. Savings accounts are a familiar example of the principles involved. One dollar invested or placed in a savings account at 15% compounded annually would be worth $4.05 in 10 years. Conversely, $4.05 received 10 years from now is worth only $1.00 today at a 15% compounded annual interest rate.

Determining present value of future dollars is called discounting, or compounding in reverse. The present worth of a dollar at some future time is the reciprocal of the future value of a dollar invested today for the same length of time, at the same rate, and with the same compounding interval.

Petroleum engineers use this concept to determine the most economical production method for a given well. From an operating standpoint, and understanding of this concept helps to realize the cost of downtime and the importance of correcting problems that impede maximum production. If a well in a field with a life expectancy of 10 years is off production, losses may not be recovered for 10 years. At an interest rate of 15% compounded annually and with constant oil prices, this production is worth only one-fourth of what it would be if it were produced today. One of the most important factors in obtaining the maximum economic gain from a well is to minimize downtime and lost production.

GAS LIFT

In wells that have too little reservoir pressure or solution gas to support natural flow, an artificial method called gas lift may be used to induce fluid flow (Fig. 10-3). There are many variations in the design of a gas lift system, but the basic concept is to take gas from an external source and inject it into the produced fluids within the tubing string. This decreases the weight of the fluid column and permits the well to flow.

Fig. 10-3

Gas lift installation. Gas in injected down the well through the space between the casing string and the tubing, enters the tubing through the operating gas lift valve, and inside the tubing mixes with the oil and lifts it to the surface.

In operation, gas under pressure is injected into the space between the casing and the tubing and enters the tubing through an open gas lift valve. The liquid in the tubing above the valve is displaced and/or lightened when the gas mixes with it, and it can rise to the surface along with the expanding gas. Once the gas and liquid reach the surface, the gas is separated from the oil. Then it is compressed once again to a high pressure and reinjected into the casing/tubing annulus to begin the cycle again.

As long as the gas is injected at a more or less constant rate, the system is classified as "continuous" gas lift. Eventually, though, the reservoir pressure will decline to a point where, even with the assistance of the supplemental gas lift supply, it will not support the flow of oil. At this stage, one of the "intermittent" gas lift systems may be used. In this technique, more time is given for liquid to build up in the tubing. Then the gas is injected into the well at predetermined time intervals and displaces the fluid to the surface in slugs.

A special type of gas lift is the plunger lift system for wells producing small amounts of fluids. An accumulator chamber for well fluids is installed at the lower end of the tubing. When enough fluid has accumulated, a plunger pushes these fluids to the surface. Power for forcing the plunger to the surface is supplied by high-pressure gas. When the plunger reaches the surface, the high-pressure gas below the plunger is released, and the plunger drops back to the bottom of the tubing until its next trip to the surface.

Gas lift is widely used as an artificial lift technique in offshore operations. Continuous flow gas lift is the preferred method of gas lifting offshore wells because the high- and low-pressure piping systems are usually of limited capacity. There are many onshore gas lift installations as well.

During the early part of the 1800s, water wells were often produced with the aid of air lift. This was accomplished with the injection of compressed air into a small string of tubing in the well in order to lift water from the bottom of the well. The same principle was later applied to oil wells, but natural gas replaced air as the injection medium to reduce corrosion and fire hazards.

CONTINUOUS FLOW

Using open-ended tubing is the simplest form of gas lift. Fig. 10-4 shows a water well under static conditions. Because the formation pressure is too low to force the water to the surface, some form of artificial lift must be applied. Please note that the hydrostatic head of the fluid in the well is equal to the pressure in the rock formation.

Production is begun by injected air or gas into the tubing which mixes with the fluid above the bottom of the tubing to lower the fluid gradient, causing the well to produce. This type of lift is known as continuous gas lift. A continuous-flow system is often installed in a well long before it stops producing in

Fig. 10-4
Open end tubing gas lift of a water well.

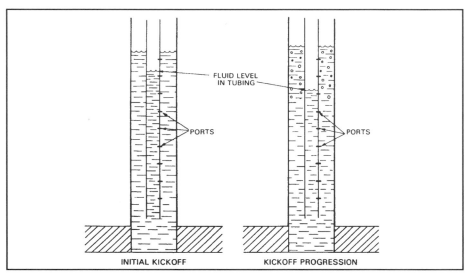

Fig. 10-5
Open end tubing with kickoff ports.

127

order to increase production and to keep the well from stopping flow. In most cases, the gas is injected into and down the annulus and the well produced through the tubing. For larger volumes, however, the gas may be injected down the tubing and the well produced through the annulus.

With intermittent flow (Fig. 10-7), injection of gas is terminated periodically, allowing fluid to enter the tubing until it reaches the desired level above the lowest gas-lift valve. The only back pressure on the producing formation is the hydrostatic pressure of the gas above the fluid in the tubing and the hydrostatic pressure from the relatively small column of fluid in the tubing, which is very small. The sudden injection of gas through a large port in the lower gas lift valve results in the rapid expulsion of the accumulated fluid as a slug, with reduced slippage of gas upward through the fluid. With proper design and control, this type of gas lift installation can be very efficient and can be used to produce a well with fairly low bottom hole pressure.

In some wells with very low bottomhole formation pressures and high productivity indexes, (b/d per psi, drawdown) a specialized form of gas lift is used known as chamber lift. This system functions in the same way as other types of intermittent gas lift systems except that when the injection gas is off, incoming well fluids accumulate in a chamber having a larger diameter than the tubing. For the same volume of fluid produced, the hydrostatic head and producing bottom hole pressure are reduced. The use of the chamber lift achieves a lower flowing bottom hole formation pressure than either continuous flow or intermittent flow gas lift. The kickoff valves above the chamber are the same as those on regular intermittent gas lift installations (Fig. 10-8).

As the gas enters the annulus from the tubing, the density of the fluid in the annulus above the point of injection is decreased. The pressure required for gas injection and the hydrostatic pressure at the bottom of the well decreases. Since the formation pressure is now greater than the hydrostatic pressure at the well bottom, fluids flow into the wellbore. The gas bubbles created by the injection at the bottom of the tubing expand as they rise in the annulus, doubling in size each time the hydrostatic head above them is halved.

This form of gas lift has worked satisfactorily on shallow wells as on wells with high bottom hole pressures. However, on deeper wells, the pressure required to kick off the gas lift is excessive. To remedy this condition, small holes, or ports, have sometimes been drilled in the tubing (or perforated) at intervals from the top of the static fluid level to the bottom of the tubing. This is illustrated in Fig. 10-5.

In this arrangement, a much lower kickoff pressure is required. However, when the injection point progresses to the bottom of the tubing, gas continues to be injected at the higher open ports, decreasing the efficiency of the lift substantially. For this reason, gas lift valves have been developed so that the points of injection during unloading can be designed to close once the operating injection depth has been read (Fig 10-6).

Fig. 10-6
Schematic of gas-lift valve installation.

Fig. 10-7
Intermittent-flow gas lift.

Fig. 10-8
Two types of chamber gas lift.

INTERMITTENT FLOW

The development of gas lift technology has progressed rapidly since 1930, leading to the development of intermittent flow gas lift. Figs. 10-5 and 10-6 show continuous injection of gas for lifting of fluids. This type of injection places a serious limitation on wells with low bottom hole formation pressures, because of the continual application of pressure from the gas being injected.

Gas lift can also introduce certain problems into the production process. Under continuous flow gas lift, relatively high back pressure is maintained on the formation. The use of high pressure gas can be a safety hazard. The effects of injection gas pressure on casing and the formation of hydrates in surface lines can become problems with gas lift.

PLUNGER LIFT

Plunger lift is the least used of all artificial lift methods. Fewer than one percent of all artificially lifted wells rely upon it. It is used most frequently in situations where some fluids are already flowing. In some wells, however, it is particularly applicable, especially in high gas-liquid ratio oil wells or gas wells with low bottom hole pressure and low productivity. In those wells, tubing flow velocities are too low to carry the liquids from the well. Separation takes place in the tubing: the well loads up with liquid and ceases to flow. Plunger lift utilizes a plunger that travels up and down inside the tubing. Internally the plunger contains a bypass valve that is opened when it hits the top of the tubing and is closed when it hits the bottom (see Fig. 10-9). The plunger's fit with the tubing reduces liquid slippage back through the gas that is propelling it as shown in Fig. 10-10.

Fig. 10-9
Main components of plunger lift.

Fig. 10-10
Plunger lift cycle showing: (A) well closed with little surface pressure, the plunger being held by the catcher, and the bypass valve open; (B) well closed in, pressure building, plunger released with bypass open, and fluid accumulating in the bottom of the tubing; (C) well closed in, plunger hitting bottom and bypass valve closing with fluid on top of plunger; (D) well opened, plunger with fluid load being lifted upward by expanding gas, and plunger bypass valve closed; (E) well open, plunger hits top, catcher holds, and bypass valve is opened. The well is then closed, and the cycle repeated.

Plunger lift is used to prolong the life of oil and gas wells flowing on energy from the well. However, with a packer, standing valve, and intermittent gas lift installation, plunger lift can utilize an external gas power source with better results than can be achieved with intermittent gas lift alone. Plunger lift is also used on wells where production is hampered by deposits of paraffin, salt, or scale in the tubing. Plunger action in the tubing helps to remove these deposits before they become bad enough to restrict production.

ADVANTAGES AND DISADVANTAGES

Gas lift has many advantages as a method of artificial lift where it can be applied. It is relatively simple to operate, the equipment required is relatively inexpensive, and it is flexible. Both high volumes and low volumes can be produced, whether continuous flow, intermittent flow, or chamber lift. It has proven effective under adverse well conditions; sand and other solids can be handled with few problems. Corrosion and high gas-oil ratio production can be handled more effectively than in other artificial lift methods and can be also used effectively in crooked holes. Gas lift can also be designed for use with wireline systems. With wireline, bottomhole pressure measurements can be obtained with little difficulty.

Among other advantages of using gas lift are its low operating costs and lower failure rate in some applications when compared to other systems. Gas lift is unobtrusive when used in urban sites, and requires less space; it can be applied to offshore production platforms.

There are also some disadvantages that also must be considered before selecting a gas lift installation. A source of high pressure gas must be available, and compression can greatly increase initial capital expenditures. Depending on market values, make-up gas for losses to a closed, compressed surface system can also be expensive. The use of gas lift for one-well leases or small fields is not usually cost effective. Gas lift is a poor choice for producing deep wells where high drawdowns, or low bottomhole pressures are present. Intermittent systems are particularly inefficient in those cases. Accurate gas measurements are not easily obtained, and surging flow can be a source of operating problems with surface equipment.

Plunger lift systems are operated automatically or with pressure controllers in most cases, while a few cases of manual installations may be found. The automatic time cycle controllers, plungers, and catchers used on plunger lift systems may vary widely.

The cost is the greatest advantage in using plunger lift. The plunger lift systems are relatively inexpensive to install and operating costs are low when compared to other systems. The plunger lift systems can be installed by wireline, and, in the case of offshore wells, no extra platform space is required.

Plunger lift can also be adapted to directionally drilled wells and can be used on wells that are already operating with intermittent gas lift, improving the performance and efficiency. Plunger lift installations are unobtrusive in urban locations, and their use has been proven effective in preventing paraffin and scale buildup in tubing.

The greatest disadvantage of plunger lift is its unsuitability for use in high-volume, high productivity wells. Plunger sticking and problems with sand production pose potential interruptions in production. Another disadvantage in plunger lift is that surging flow from a well may tend to hamper the effectiveness of the surface equipment assembly.

SUCKER ROD PUMPING

By far, sucker rod pumping is the most widespread form of artificial lift. Used since the earliest days of the oil industry, sucker rod pumps are functionally the same as water well pumps used as long ago as 1,500 years in China, Egypt, and Rome. The basic components of a sucker rod pumping system are: the subsurface pump, sucker rods to transmit power from the surface to the pump, and a surface pumping unit to furnish power to the sucker rods in the form of a reciprocating motion. The beam pumping unit illustrated in Fig. 10-11 is the most widely used type.

OPERATION

The subsurface pump, in its simplest form, consists of a plunger working up and down in a closely fitted barrel. The plunger contains a check valve that permits fluid flow to flow up but not down. The check valve, also called a traveling valve, is usually a ball-and-seat type in today's pumps. A second type, called a standing valve, is a ball-type valve located at the bottom of the barrel which, like the check valve, allows fluid to flow up but not down (Fig. 10-12).

The principle of the simple sucker rod pump is illustrated in Fig. 10-13. To start, the plunger is stationary at the bottom of the stroke. Both the traveling valve and the standing valve are closed at this time. The fluid column imposes hydrostatic pressure in the tubing string above the standing valve. The load on the polished rod, the top rod in the sucker rod string, and on the pumping unit is the weight of only the rod string. As the plunger move upward, the traveling valve remains closed, and the load of the fluid in the tubing is picked up

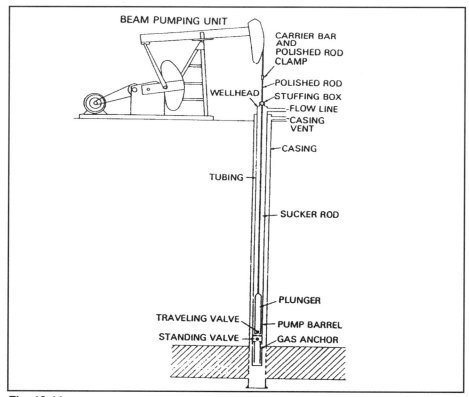

Fig. 10-11
Basic components of a sucker rod pumping system.

by the rod string. The load of the rod string and that of the fluid column. With minimal leakage between the plunger and the pump barrel, the pressure between the traveling valve and the standing valve is reduced, so the standing valve opens, allowing fluid to flow from the wellbore into the pump barrel. At the top of the stroke, the plunger is stationary and both valves are again closed, so the fluid load is still being held by the plunger and the traveling valve. Assuming that the pump barrel is now filled with fluid and that the fluid is incompressible, the traveling valve will open as the plunger starts downward. The weight of the fluid column in the tubing will be transferred to the standing valve and the tubing, and the load on the polished rod and the pumping unit again will be only the weight of the rods. Further downward movement of the plunger will cause fluid to flow from the pump barrel through the traveling valve. Return of the plunger to the bottom of the stroke will complete the cycle.

In practice, such a polished rod loading will never occur. Inertia affects loading, pump operation is not 100% efficient, friction would alter the loads, rods would stretch when loaded, and the dynamics involved will cause variations. A near approach to this type of polished rod loading will occur, however, in pumping a single phase fluid from a very shallow well with long, slow

Fig. 10-12
Schematic drawing of a simple sucker rod pump.

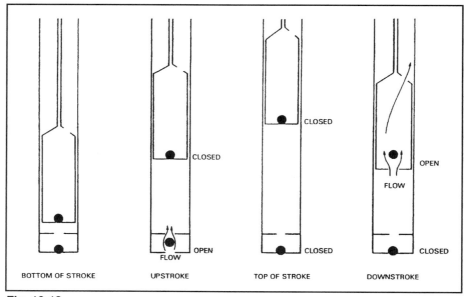

Fig. 10-13
Schematic drawing of the operating cycle of a sucker rod pump.

pump strokes. Actual rod load diagrams used to evaluate pumping performance are called dynamometer cards.

SUBSURFACE PUMPS

There are two basic types of subsurface pumps used for sucker rod pumping (Fig. l0-14). The first type is called a tubing pump and is so named because the pump barrel is run on the tubing string. The plunger is run into the well on the sucker rods. The inside diameter of a tubing pump barrel is just slightly smaller than that of the tubing on which it is run, giving the highest pumping rate for a given installation. Replacing the barrel of a tubing pump requires pulling the tubing string.

The second type of submersible pump is called an insert pump, which is run in and pulled from the tubing on the rods. The insert pump was developed prior to 1870 but not used extensively until after about 1920. Since the insert pump can be pulled as a single unit, it is preferred to the tubing pump for use on deeper wells.

The most common tubing and insert pump designs are controlled under API specifications. Certain non-API pump designs such as the casing pump and the multistage pump, have shown to be effective under certain special well conditions.

Fig. 10-14
Two types of sucker rod pumps. The pump at left is a tubing type pump. On the right is a stationary barrel, top hold-down rod type. The rod-type pump and barrel can be removed from the well without removing the tubing.

SUCKER ROD PUMPS

The earliest sucker rods were made of wood, usually hickory, with metal end pieces. Iron and steel rods came into use about 1880-90 and became common by 1900. API standards for sucker rods were first adopted in 1927.

Continued improvements in the metallurgy used in sucker rod manufacture have increased the strength and load capacity of sucker rods. However, even with these improvements and the use of tapered rod strings, the maximum practical depth of sucker rod installations is about 10,000 ft, with a few low volume installations as deep as 13,000 ft. Lighter, stronger materials are necessary for installations below 15,000 ft; a maximum depth of 20,000 ft is possible with current technology. The use of state of the art fiberglass material is currently being explored for use in sucker rods for special applications.

BEAM PUMPING UNITS

The beam pumping unit furnished power to the top of the sucker rod string in the form of reciprocating motion. The length of the stroke may vary from less than 1 ft to as much as 80 ft. The earliest beam pumping unit installations utilized a cable-tool rig after the drilling process was completed, using the walking beam from the rig to operate the subsurface pump. The structural members of these units were wooden with metal bearings and hardware. Power was furnished from steam or single-cylinder low speed internal combustion engines, transmitted through a belt drive. Electric motor drives were sometimes added later. In these installations, the derrick was left over the well, and the power unit and bull wheel were used for well servicing operations. The same surface equipment was used for drilling, production, and servicing. These installations, with innovations, were used until about 1930. By that time, deeper wells were being drilled, pumping loads had become heavier, and the cable-tool rig as a pump had outlived its efficiency. Fig. 10-15 illustrates the early beam pumping unit derived from a cable-tool rig.

The modern beam pumping unit was largely developed during the 1920s, shown in Fig. 10-16. The development of efficient mobile well servicing units had eliminated the need for integral power hoists at each well, and the development of durable, efficient gear reduction units paved the way for the production of beam pumping units with higher speed and prime movers with lighter weight.

The relatively high rotating speed of the prime mover is first reduced by the belt drive and then by the gear reducer to rotate the crank at the desired strokes-per-minute rate. Rotation of the crank is converted through the crank arm, crank pin bearing, itman, equalizer, and equalizer motion is then converted to linear motion of the polished rod by the horsehead and the hanger. With proper unit alignment, this motion should not impose any bending

Fig. 10-15
Early beam pumping unit.

moment on the polished rod. The polished rod and the stuffing box provide a seal between the rods and the tubing at the surface to divert the fluid being pumped into the flow line.

Counterbalance

The counterweight shown on the crank arm of the beam pumping unit, Fig. 10-15, is an important component of the system. The counterbalance weight can also be placed on the walking beam, or an air cylinder can be used for the same purpose (Fig. 10-16). Pumping units can be described as beam balanced, crank balanced, or air-balanced (Fig. 10-17).

The purpose of counterbalancing can be visualized by studying the motion rod string and the pumping unit in conjunction with the idealized pump operation illustrated in Fig. 10-13. In this simplified situation, the load on the polished rod during the upstroke is the weight of the rods plus the weight of the fluid. During the downstroke, it is the weight of the rods only. Without any counterbalance, the load on the gear reducer and prime mover during the upstroke goes in one direction. During the downstroke, the load is in the opposite direction. This type of loading is highly undesirable. It causes unnecessary wear and tear and fuel/power consumption. In practice, a counterweight equivalent to the rod weight plus approximately one-half the fluid weight is used. Proper counterbalance results in the lowest loads possible on the gear box and prime mover, reduced failures and downtime, and reduced fuel or power requirements. It has been estimated that as many as 25% of the beam pumping units in operation are not properly balanced.

OTHER RECIPROCAL PUMPING UNITS

The beam pumping unit is used on more than 99% of all wells on sucker rod pumps, but a number of different types of sucker rod pumping units that have limited application are available. Theoretically, any device that will pro-

Fig. 10-16
Modern beam pumping unit.

vide an up-and-down motion to the sucker rod string could be used. Hydraulic and pneumatic pumping units using high-pressure hydraulic and pneumatic cylinders to operate the sucker rod string have been in use since about 1900.

Pneumatic Units

With pneumatic pumping units, one or more cylinders are placed over the well-head, and a counterbalance system using compressed air or gas is provided (Fig. 10-18). In such a system, high pressure gas is applied to the underside of the pistons of the inside power cylinders on the upstroke. The low-pressure gas above the pistons is discharged to the sales line. On the down stroke, high pressure gas is applied to the top of the pistons, and the low pressure gas below the pistons is discharged to the sales line. Pressure is applied from the counterbalance tank to the two outside counterbalance cylinders, balancing the rod weight and one-half the fluid load, as in a beam pumping unit. The top side of the counterbalance cylinders is vented to the atmosphere. Pressure on the power piston at the top and bottom of the stroke is reversed by a spool valve operated by a reversing rod attached to the piston rod harness. Units of this type are particularly applicable to pumping water from gas wells where well-head pressure is significantly greater than sales line pressure, providing operating power. Also, in most installations, these units have no power costs since

Fig. 10-17
Types of counterbalancing beam pumping units: (A) beam balance unit; (B) crank balanced unit; and (C) air balanced unit.

they use available gas pressure drop from casing head to sales line as a source of energy.

Hydraulic Units

In the late 1940s, hydraulic sucker rod pumping units with strokes longer than 20 feet were in use (Fig. 10-19). Compared to the 10 ft maximum stroke lengths of earlier beam pumping units at the time, these newer units provided smoother operation, fewer rod reversals, and increased compression ratios to reduce gas locking. Initial costs were high, however, and now that beam pumping units with 25 ft strokes are in use, the hydraulic long stroke pumping units have lost favor, and few are still in operation.

Cable Wind and Wheel Jack Units

Many other types of sucker rod pumping units have been used to a limited extent. The cable-wind pumping unit (Fig. 10-20) can be used to achieve 40-

Fig. 10-18
Pneumatic sucker rod pumping unit.

80 ft stroke lengths, reducing dynamic loads, rod reversals, gas locking, and power consumption. This system can be applied to deep pump applications. The wheel jack, on the other hand has been applied to shallow pumping applications only.

In spite of the many variations in sucker rod pumping units, the beam pumping unit is preferred because of its reliability, simplicity, flexibility, and familiarity. Though the basic operating principles remain the same as those of the earliest pumping units, the modern beam pumping unit system, through technical advances, has become an increasingly efficient component of a production installation.

ADVANTAGES AND DISADVANTAGES

Advantages

Sucker rod pumping is used widely enough to be familiar to most operation and maintenance personnel. It can be used over a wide range of production capabilities and is able to produce most wells to depletion at limited rates and depths. Sucker rod pumping systems are highly reliable and are relatively easy to analyze by using several different means: observation, dynamometer, and well sounding.

High temperature or viscous oil can be produced, and corrosion and scale problems are easily treated. Sucker rod pumps can be driven by either electrical power or fuel gas, the electric drive being easily adaptable to time-cycle or pump-off control. Finally, the salvage value of a sucker rod pumping unit is an added benefit in maintaining a low cost production operation.

Disadvantages

Among the disadvantages of sucker rod pumping systems are: These installations are not suitable for crooked hole work; depth and volume for which they can be used are limited by rod weight and strength considerations; and high gas-oil ratio wells, as well as sand and paraffin content in production fluids, further impair the effectiveness of the units.

Certain physical characteristics of the units may also discourage their use. Their high profile is unsightly in urban locations and interferes with operation of center-pivot irrigation machines on farmland. Their overall weight and size

Fig. 10-19
Hydraulic long strike sucker rod pumping unit.

Fig. 10-20
One type of cable pumping unit.

can prohibit their use in offshore installations. For servicing downhole equipment, the added inconvenience of using a pulling unit must be considered.

ROTARY UNITS

A relatively new pumping system to the oil industry is the progressive cavity pumping system (Fig. 10-21). This system consists of a rotor constructed of chrome plated steel in the form of an external helix. The stator in which the rotor turns is a synthetic elastometer with a double internal helix permanently bonded into a steel housing. Rotation of the shaft by a vertical spindle electric motor at the surface causes the shaft to stretch a predetermined amount to maintain the shaft in tension and causes a cavity containing well fluid to rise (Fig. 10-22).

The advantage of this pumping system over reciprocating beam pumping systems is that its power costs are lower, it pumps viscous fluids more readily, it handles sandy fluid better, and its low profile at the surface makes it more desirable for agricultural and urban installations.

The disadvantages of this pumping system are that the depth is limited to about 4,000 ft by the temperature resistance of the elastometers, its capacity limit is about 400 b/d of fluid with state of the art technology, and it is little known within the industry.

Fig. 10-21
Progressive cavity pumping unit.

Fig. 10-22
Cavity pumping unit.

HYDRAULIC PUMPING

Hydraulic pumping is relatively new when compared with sucker rod pumping. Hydraulic pumping was introduced to the industry by C.J. Coverly (Kobe Inc.) in the early 1930s. The subsurface pump used in this system is similar to the pump used in sucker rod pumping but is operated by a directly coupled hydraulic engine powered by high-pressure oil (as much as 5,000 psi.) that is pumped from the surface.

Kobe Pumping Units

The earliest hydraulic pumping systems were the insert type, run inside the production tubing on a small diameter string of power-oil tubing, Fig. 10-23. In this case, the hydraulic pumping unit is landed in the bottom of the production tubing in a manner similar to the landing of a sucker rod insert pump. Discharge from the hydraulic engine mixes with the well fluids being pumped and flows up through the production tubing. Since the hydrostatic head of power oil in the power-oil tubing is equal to the hydrostatic head of the oil in the production string, the only power being used is that required to pump the well fluids and to overcome friction. Any gas being produced can be vented through the casing and, as with a sucker rod pump installation, a gas anchor can be installed below the pump to aid in gas separation. The pump size and the engine size limit the production rate to the rate of the pump that can be run inside the production tubing.

By using a conventional casing installation, larger hydraulic pumping units can be run and larger volumes pumped for a given casing size (Fig. 10-24). In this installation, power oil and production fluids are pumped down the power oil tubing to the down hole hydraulic pump and mixed with new production fluids which are pumped to the surface up the casing string.

A disadvantage of this installation is that produced gas must pass through the pump, and the casing is thus exposed to produced fluids. With both the conventional casing and conventional tubing pump installations, the power-oil tubing must be pulled from the well in order to service the pumping unit.

In 1950, a hydraulic pumping unit was introduced that eliminated the need for pulling the power-oil tubing in order to retrieve the pumping unit. These so-called free pump installations are pumped down by hydraulic pressure applied in the same manner as with pumping and removed by reversing the flow of hydraulic power oil (Fig. 10-25). The engine and the pump of this unit are essentially the same as in the conventional hydraulic pumping unit installation. However, the landing nipple and the bottom seating plug are different, and the free pump has swab nose assembly equipped with two swab cups at the top of the unit. These cups are normally oriented to provide a pressure seal during retrieval. In some deep or crooked holes the top cup is

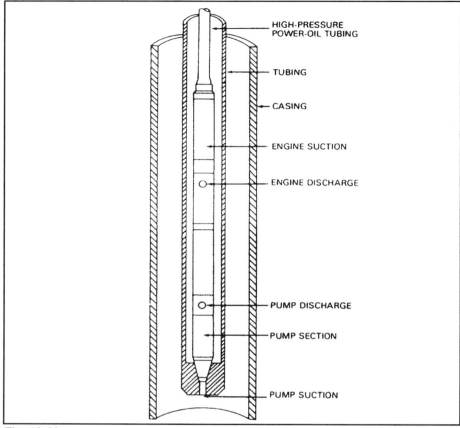

Fig. 10-23
Insert-type hydraulic pumping system.

reversed so that a pressure seal is provided for retrieval. The swab nose has a fishing neck for wireline retrieval if the unit cannot be pumped out in the normal manner.

BOTTOMHOLE PUMPING UNITS

Since the Kobe hydraulic pumping unit was introduced, a number of companies have developed hydraulic pumping units with a wide variety of design features. All have the following components:

• A pump piston reciprocating in a cylinder.
• Standing and traveling valves for the pump.

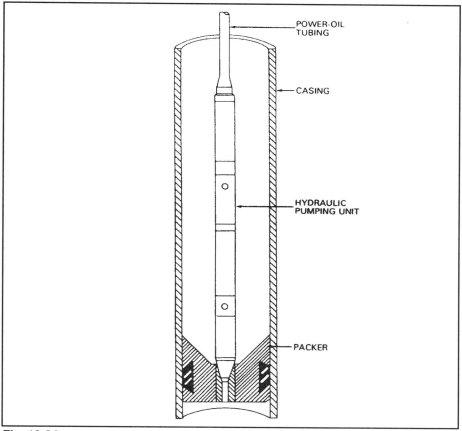

Fig. 10-24
Casing hydraulic pumping system.

- An engine piston directly coupled to the pump piston.
- A reversing valve to reverse the application of hydraulic power-oil.

One design has one single-acting pump piston that pumps only on the upstroke and one engine piston (Fig. 10-26). Other designs have a double-acting pump that smoothes out the power demand, with tandem pumps for increased volume, compound pumps for increased compression ratio, and tandem engine pistons for increased power. A variety of components for varying bottomhole installations are also available to adapt these units to well conditions (Fig. 10-27).

PARALLEL FREE PUMP

CASING FREE PUMP

Fig. 10-25
Free hydraulic pumping system.

SURFACE TREATMENT AND PUMPING EQUIPMENT
Power Oil Systems
OPEN Systems

Most of the hydraulic pumping systems now in use have an open power-oil system which uses produced crude oil as the hydraulic fluid (Fig. 10-28). Wellhead connections are available for the most common types of downhole installations. The power-oil tank has a gas boot and spreader to reduce agitation and promote solids settling. Intake for the power-oil plunger pump is near the top of the oil in the tank to ensure pumping of the cleanest possible oil. Commonly accepted standards for maximum solids contamination of power-oil for a hydraulic pumping system are as follows:

- Not more than 20 ppm total solids.
- Not more than 12 lb of salt per 1,000 bbl of oil.
- No solids particles larger than 15 microns (in small quantity).

CASING

TUBING

ENGINE PISTON

ENGINE BARREL

MIDDLE ROD

SEATING CUPS

PUMP PISTON

PUMP BARREL

PACKER

CHECK VALVE

SWAB

PUSH ROD

REVERSING VALVE

HARNESS

SOFT PACKING

TRAVELING VALVE

PUMP STANDING VALVE

TUBING STANDING VALVE

UPSTROKE

DOWNSTROKE

Fig. 10-26
Another type of free hydraulic pumping system.

151

Fig. 10-27
Examples of bottomhole hydraulic pumping assemblies.

Closed system

When emulsion treatment is a problem, and clean power oil cannot be easily maintained with an open system, a closed power-oil system can be used. This closed system recirculates a clean fluid, oil or treated water. A closed power-oil system requires one more string of tubing in the well than similar open power oil installations. Typical power-oil systems are shown in Fig. 10-29.

Power Oil Pumps

The high-pressure positive-displacement power-oil pumps used to furnish power for hydraulic pumping are usually equipped with metal-to-metal pistons

Fig. 10-28
Open hydraulic power-oil system for hydraulic pumping.

and cylinder liners that can be sized to accommodate the volume and pressure requirements of the system. These pumps are usually used to furnish power oil to several wells. One manufacturer furnishes five different sizes of power-oil pumps, ranging from 30 to 250 hp input, that will operate at pressures of as much as 6,000 psi. As many as seven different piton-liner combinations are available for each pump.

Hydraulic Jet Pumping Units

A specialized form of hydraulic pump that is gaining wide acceptance is the hydraulic jet pump. It usually is run as a casing- free pump, which requires only one string of tubing and no gas venting (Fig. 10-30).

The main working parts of the hydraulic jet pump are nozzle, throat and diffuser. The nozzle converts the high-pressure, low velocity energy of the power fluid to high velocity, low pressure energy. The power fluid is then mixed with the low pressure pump intake fluid in the throat to produce a low pressure stream with a velocity less than that of the nozzle exit but a high velocity nonetheless.

The velocity energy of this mixed stream is then converted to static pressure in the diffuser to provide the necessary to lift fluid from the well (Fig. 10-31). The power fluid may be either oil or water.

The jet pump is sensitive to the mass rather than the volume of the fluid being pumped. This enables it to handle some but not a lot of gas. A hydraulic jet pump is also sensitive to variations in intake pressure, which, at the desired rate of production, must be known in order to select an efficient nozzle and throat combination. The relationship between pump depth, intake pressure, production rate, power-fluid rate, and power-fluid pressure is of such complexity that nozzle and throat sizing is usually done by computer.

ADVANTAGES AND DISADVANTAGES

Advantages

Among the advantages of hydraulic pumping are its adaptability to situations in which hole deviation occurs, its flexibility under changing well conditions, and its efficiency on multiple well installations. Hydraulic pumps are not limited to the depth and volume restrictions encountered in using sucker rod pumps.

Open hydraulic pumping units are also easier to install and less obtrusive when installed: their low profile allows for their use in urban areas; they can be powered by electricity or gas; and, with free-pump installations, no pulling unit is required to service or exchange downhole equipment.

Fig. 10-29
Bottomhole hydraulic pumping assemblies for closed power-oil systems.

POWER FLUID

NOZZLE

THROAT

PRODUCED FLUID

DIFFUSER

PUMP DISCHARGE

PUMPED FLUID

PRESSURE BOMB CHAMBER

PUMPED FLUID

Fig. 10-30
Casing free hydraulic jet pump installation.

Disadvantages

The use of hydraulic pumping units is not without drawbacks. Produced fluids with high sand content or other abrasive materials may cause difficulties. Likewise, corrosion may require the use of a closed power-oil system, which is an added cost. High-pressure flammable power oil can be a safety problem, and the production of high gas-oil ration fluids may even require an added string of tubing. Cavitation, creation of a partial vacuum, must be avoided in the design of hydraulic pumps.

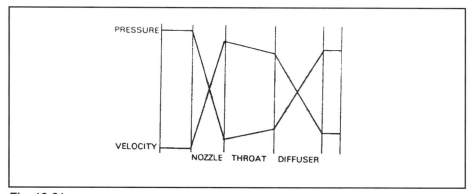

Fig. 10-31
Pressure and velocity profiles through a hydraulic jet pump.

Initial costs for a hydraulic pumping installation tend to be high, particularly since operation and maintenance personnel must be specially trained. These costs and the fact that failure rates can be about twice as high as those of sucker rod pumping suggest that hydraulic pumping is not always the best alternative.

ELECTRIC SUBMERSIBLE PUMPS
The electric submersible multistage centrifugal pump was introduced to the oil industry as a means of artificial lift by the Reda Pump Company in the late 1920s. Since that time several other companies have developed electric submersible pumps for oil field use. There are a large number of sizes, capacities, and voltages now available for this type of pump. In a conventional installation, the pump assembly and electric motor are run into the well on the production tubing. Electric power is conducted to the assembly by a cable attached to the tubing (Fig. 10-32).

Table 10-1
Maximum Horsepower for Various Casing Sizes

Casing Size (OD)	Maximum Hp (60 Hz)
4-1/2"	127.5
5-1/2"	240
7"	600
8-5/8"	720

Bottomhole Assembly

In a conventional electric submersible pump installation, the bottomhole assembly consists of an electric motor, a multistage centrifugal pump, and a cable harness (Fig. 10-33). The electric motor is a three-phase induction type that is oil-filled for cooling and lubrication. Cooling is by heat transfer to the well fluid, which flows up around the motor to the gas separator and the pump suction when the assembly is above the producing zone. These motors normally operate at 3,500 rpm on a 60-cycle power supply or 2,900 rpm on a 50-cycle power supply. They are available in many sizes, ranging from a few horsepower to more than 700 hp. The maximum size available for a given size casing being shown in Table 10-1. A number of motor voltages are available from about 300V to more than 3,000V for the most economical selection of motor, cable, and switchgear combinations.

The protector section is designed for pressure equalization between the oil fill in the motor housing and the wellbore to allow for thermal expansion and contraction of the oil. Mechanical seals prevent migration of well fluids into the motor housing or oil out of the motor housing.

The electric cable used is specifically designed for oilwell application in copper wire sizes from No. 1 through No. 8. The size selected is based on amperage and voltage drop, with voltage drop usually limited to 30 V/1,000 ft. A standard cable made of coated copper conductors, polypropylene-ethylene insulation, nitrile jacket, and galvanized steel or monel armor has a temperature limitation of 205° F. at 1,500 psi. Increases in pressure reduce the temperature limitation. More expensive cables with ratings to 400° F. at 1,500 psi are available. The standard cable is considered to have a 10 year lift at 167° F., with each 16° F. increase in temperature reducing the expected life by half. I5 is the cable temperature limitation that limits the setting depth of electric submersible pumps to about 10,000 ft. Cables are usually round, but flat cables are available for applications where clearances dictate their use.

Surface Equipment

A typical installation of surface equipment for a submersible pumping system consists of a wellhead to support the tubing and downhole equipment and provide a seal around the tubing and electric power cable, a junction box to provide a weatherproof connection between the downhole and the surface cable, switchgear, and a three-phase transformer bank (Fig. 10-34). The switchgear is tailored to the individual situation. The gear is weatherproof and may have only a magnetic starter and overload protection, recording ammeret, signal lights, time-cycle controllers, or automatic remote control instruments. Power surge protection may also be provided.

Fig. 10-32
Electric submersible bottomhole pumping unit.

Fig. 10-33
Cutaway drawing of submersible pumping unit.

The transformer bank is designed to convert the distribution system voltage to the required subsurface voltage. The transformers may be either step-up or step-down and are usually equipped with lightening arrestors and capacitors to prevent motor damage from voltage surges.

Application

Electric submersible pumping is usually used for high volume high water cut wells that perform much like water wells. Most installations are for production rates in excess of 1,000 b/d for the following reasons:

- More horsepower can be delivered to the pump in oil well casing than with other artificial systems.
- Centrifugal pumps are capable of much higher rates than positive displacement pumps in oil well casing.
- Other systems are usually more economical at lower production rates.

The conventional installation, shown in Fig. 10-35, is applicable only where

Fig. 10-34
Surface equipment installation for electric submersible pumping unit.

the pump can be set above the producing interval. If this same pump configuration, Fig. 10-36, were used in an installation where the pump had to be set below the producing zone, there would be no fluid around the motor to provide cooling. An installation with the pump below the producing interval might be required to reduce wellbore pressure to a low enough level to provide the desired inflow or to ensure maximum bottom gas separation in high gas-oil ratio wells. Fluid flow around the motor in this setting can be obtained by placing a shroud around the pump suction, protector, and motor sections of the bottomhole assembly. Except for the shroud, the overall pumping assembly is the same as in the conventional installation. Shrouded pumps cannot be used in casing smaller than 5-1/2 in. OD. The gas separator section attached to the pump suction acts to separate a substantial portion of any free gas from the well fluids.

As with other artificial lift systems, provision must be made for venting of this gas from the casing. Also, as with other systems, the separation of gas from liquid in the small space available is not 100% efficient. Multistage centrifugal pumps with open impeller inlets are not highly susceptible to gas locking. However, the injection of gas reduces overall pump efficiency and increases pumping expenses.

Depending on flow and head requirements, the pump suction of an electric submersible pump installation may consist of any number from one to hundreds of centrifugal pump stages. Usually, when compared to positive-displacement pumps, a single stage centrifugal pump will provide high flow rates at much lower discharge pressures. Since the amount of head generated by a single-stage pump depend upon the diameter of the impeller, the head per stage in a bottom hole centrifugal pump for smaller casing sizes is very limited. Lifting fluids from greater depths requires several stages. This type of pump is about 40% efficient in the smaller sizes and almost 80% efficient in the larger sizes. Pumps are available in corrosion-resistant materials for all normally encountered fluids.

ELECTRIC CABLE

A bottom intake configuration is used where casing size prohibits the desired production volume because of tubing friction loss or pump diameter restrictions. In this configuration, the pump and motor sections are reversed, with the pump intake through a stinger at the bottom of the unit. This type of installation is most commonly used in wells with 4-1/2 in. OD or heavy-wall 5-1/2 in. casing.

The high volume capability and the declining flow rate with increasing back pressure make the submersible pump applicable in many water flood projects for closed system production-injection (Fig. 10-37). Most reservoirs accept

Fig. 10-35
Shrouded submersible pumping unit.

Fig. 10-36
Submersible pumping unit with a packer.

Fig. 10-37
Production -injection system for a waterflood enhanced oil recovery operation.

floodwater at high rates and low injection pressures initially. As the reservoir fills, injectivity decreases and injection pressures increase.

CABLE SUSPENDED SUBMERSIBLE PUMPS

The electric submersible pump installations describe thus far have all been those that require pulling of the tubing in order to service the bottomhole assembly. A more recent development is the cable-suspended electric submersible pump, which can be pulled or run on the power cable. It sits in a shoe at the bottom of the tubing or in a packer in tubingless completions. The bottom hole assembly is locked in position by rollers in the lockdown head that release when tension is placed on the power cable. The cable suspended pump installation was made possible by the development of power cables designed for tensile loads in excess of 100,000 lb. The armor is two layers of plow-steel wire counter-wound to prevent twisting. This type of installation is more expensive than conventional installations described previously, but has application to some special situations.

WELL TESTING

To plan operations for maximum economic recovery, all oil and gas wells are tested at regular intervals.

In oil well testing the principal criteria are the oil producing rate, the gas-oil ratio, the gravity of the crude oil, salt water production (as a percentage of total liquid production), and BS&W (basic sediment and water) content. The importance of the oil-producing rate is obvious.

The gas-oil ratio (GOR) is an important guide to the efficiency of operations. As mentioned earlier, conservation of gas normally increases ultimate oil recovery. Thus, a low GOR usually indicates efficient production methods, and high GOR often indicates inefficient methods. In fact, where "allowable" restrictions are imposed by a government regulatory agency, periodic reports of GOR tests are usually required, and the allowables of wells—the amount of hydrocarbons a particular regulatory agency allows to be produced from a given reservoir—with excessive GORs are reduced.

The gravity of the crude oil is quite important, since the selling price of crude oil is a function of its gravity. An arbitrary measure called "API gravity," established by the American Petroleum Institute, is the commonly used gravity scale. API gravity is related to specific gravity in accordance with the following formula:

$$°API = \frac{141.5 - 131.5}{\text{specific gravity}}$$

Table 10-2
Artificial Lift Comparison

ARTIFICIAL LIFT COMPARISON

	ESP	HYDRAULIC PISTON PUMP	HYDRAULIC JET PUMP	GAS LIFT	PCP
CAPITAL COST:	3	2	2	5	1
SYSTEM EFFICIENCY:	50-60%	30-40%	10-20%	5-30%	50-80%
OPERATING COSTS:	1	3	4	4	2
RELIABILITY:	3	3	3	4	4

LOW ——→ *HIGH*
1   5

Table 10-3
Artificial Lift Limitation

ARTIFICIAL LIFT LIMITATIONS

	ESP	HYDRAULIC PISTON PUMP	HYDRAULIC JET PUMP	GAS LIFT	PCP
OFFSHORE APPLICATION	GOOD LITTLE SURFACE EQUIP. NEEDED	GOOD IF SPACE FOR POWER SYSTEM	GOOD CAN USE PROD. WTR. AS POWERFLUID	EXCELLENT	GOOD IF PULLING UNIT AVAILABLE
SOLIDS HANDLING CAPABILITIES	PROD-LESS THAN 200 PPM	PROD-LESS THAN 200 PPM POWER-LESS THAN 10 PPM	PROD-UP TO 3% POWER-UP TO 200 PPM OF 25 MM	NO PROBLEMS	NO PROBLEMS UP TO 5%
TEMP. UNITS	325°F	500°F	500°F	350°F	250°F
VOLUME CAPABILITIES (TURNDOWN)	250-50,000 BPD (FAIR)	100-5,000 (GOOD)	100-15,000 (FAIR)	50-50,000 (GOOD)	50-2000 BPD (GOOD)

As this equation indicates, an oil with an API gravity of 10 degrees (usually written 10°API) has a specific gravity of 1, which is the same as the specific gravity of water.

The API gravity and the GOR can be changed by altering the operating pressure of the gas/oil separators. Increasing the operating pressure of the separators increases the API gravity, since more of the gas will remain in solution in the oil and will reduce the GOR.

The gas-oil ratio can be further decreased and the API gravity correspondingly increased if more than one gas/oil separator, in series, is used. This technique, called stage separation, is particularly effective with crude oils of relatively high API gravity (usually more than 35° API). The volume of crude oil is also increased by this method, since some of the gas is kept in the liquid state in the crude. Sometimes the ultimate oil recovery is increased by as much as 5% through installing stage separation facilities.

The amount of salt water production is important not only because of the expense of producing the salt water but because of the appreciable cost of disposing of the salt water after it has been produced. Also, removing salt water from the reservoir contributes to pressure decline, which is undesirable. For these reasons, salt water production should be minimized.

Basic sediment and water content is present in most crude oil. As the term implies, BS&W is an emulsion of oil, water, and sediment. Most crude oil purchasers specify the maximum BS&W content that they well accept, usually only a small fraction of 1%.

CHAPTER 11

Surface Equipment

If you have every driven by an oil field, you have no doubt noticed the many kinds of equipment spaced around the lease. This is called *surface equipment,* and much of it has to do with controlling the production rate of the wells and cleaning the hydrocarbons as they rise to the surface. Let's take a look at this equipment and learn about its role.

THE WELLHEAD

The wellhead is the cast or forged steel configuration of pipes at the top of a wellbore that controls the well's pressure at the surface (Fig. 11-1). It is purposely machined for a very close fit so it forms a tight seal to prevent well fluids from leaking or blowing at the surface. Some of the heaviest fittings on the wellhead are constructed to hold pressures as high as 30,000 psi. Other wellheads simply support the tubing in the well and may not be built to contain pressures that high.

The wellhead is composed of various parts, some of which are the casinghead, the tubinghead, and the christmas trees (Fig. 11-2).

The casinghead

As the well is drilled and each string of casing is run into the hole, heavy fittings must be installed at the surface for the casing to be attached to. The casinghead is the equipment that serves this function. It is equipped with slips or other gripping devices to support the weight of the casing. The entire assembly seals the casing and thus prevents the flow or escape of fluids from the well.

Gas outlets are usually provided to reduce gas pressure that may collect between or within the strings of casing. These outlets may sometimes be used when producing the well through the casing.

During drilling and workover operations, the casinghead is used to anchor the pressure control equipment. Adaptors, packoffs, and flanges are used to

Fig. 11-1
Pumping wells are equipped with a simple type of wellhead such as this one.

accommodate progressively smaller casing sizes as drilling continues and additional strings are set. This means the BOPs must be removed and reinstalled each time a new string of casing is run into the well. As more flanges and spools are installed, they form an integral part of the permanent wellhead.

The tubinghead

The tubinghead is designed for three purposes:
• To support the tubing string.
• To seal off pressures between the casing and tubing.
• To provide connections at the surface to control liquid or gas flow.

The tubinghead is supported by the casinghead. Tubingheads vary in construction, depending on pressure. To help during well servicing, many styles of tubingheads are designed to be easily disassembled and reassembled.

Adapter

Tubing hanger

Tubinghead

Resilient, hard rock
or metal seal

Full-opening heads,
spools, hangers

Dual compression,
pressure-energized
casing packoff
with bit guide

Secondary packoff

Primary packoff

Automatic
set-and-seal slips

Casing spool

Injection and bleed
ports for field repair

Positive separation
of test and annular
pressures and casing load

Fully automatic set-and-seal
casing hanger for heaviest loads

Bit guide and single compression,
pressure-energized packing

Secondary packoff

Primary packoff

Casinghead

Conventional slips

Test ports for testing
and pressure checks

Economical positive slip
suspension for maximum loads

Fig. 11-2
Wellhead components (courtesy Gray Tool Co.).

The christmas tree

Wells that are expected to have high pressures are equipped with special heavy valves and control equipment at the casinghead or tubinghead before the well is completed. These valves control the flow of oil and gas from the well and are commonly called the christmas tree.

Pressure gauges are part of the wellhead and christmas tree assembly and are used to measure the casing and tubing pressures. Knowledge of these pressures allows the operator to control the well's production better.

Sand is sometimes produced with the well fluids. The abrasive fine particles can cut valves, fittings, and chokes.

The master valve is the key to closing the well in an emergency, so it must be kept in good, dependable condition. The accepted practice is to use it only when absolutely necessary to keep it from being abraded by sand particles.

SEPARATION METHODS

Well fluids are a complex mixture of liquid hydrocarbons, gas, water, and some impurities. The water and impurities must be removed before the hydrocarbons are stored, transported, and sold. Liquid hydrocarbons and objectionable impurities must also be removed from natural gas before the gas goes to a sales line. Nearly all of the impurities cause various types of operating problems.

Natural gas, liquid hydrocarbons, and hydrocarbons are separated by various field processing methods. These methods include time, chemicals, gravity, heat, mechanical or electrical processes, and combinations of these.

Separators

A separator is a piece of equipment used to separate wellstream gas from free liquids. The size of the separator depends on the rate of natural gas flow and/or liquids going into the vessel. The operating pressure of the vessel depends on the pressure of the gas sales line, the flowing pressure of the well, and the operating pressure that the lease operator wants.

Separators are build in various designs, including vertical, horizontal, and spherical. Some separators are two-phase types, which means they divide the produced materials into crude and natural gas (Fig. 11-3). Other separators are three-phase types, which means they divide the produced materials into oil, crude, and free water. Sometimes it is more desirable to use more than one stage of separation to increase fluid recovery.

While natural gas leaving the separator no longer contains free liquids, the gas may contain significant amounts of water vapor. Water vapor in gas at high pressure can cause serious operating problems by forming gas hydrates—an ice-like sludge. When hydrates form in a gas-gathering or distribution line, total or partial blockage of the pipeline may result.

TREATMENT METHODS

Dehydrating natural gas

Several methods can help prevent hydrates from forming in a gas line:
- Heat the gas stream so that gas temperature will not drop to the level at which hydrates form.
- Add an antifreeze agent such as methanol or glycol to the gas stream.

Fig. 11-3
Vertical two-phase separator.

- Remove the water vapor by using a glycol dehydrator, which consists of a vertical pressure vessel (a glycol absorber) that allows the glycolto flow downward as the gas flows upward.
- Dehydrate using drying agents such as alumina, silica gel, silicon alumina beads, or a molecular sieve.
- Expand the gas and refrigerate using heat exchangers.

Most dehydrated gas that goes to a sales line contains no more than 7 lbs of water vapor per million cubic feet of gas (MMcf).

Other objectionable impurities that are found are hydrogen sulfide and carbon dioxide. These impurities may be removed using chemical reactions, physical solutions, or absorption. The technique used depends on how free of impurities the gas must be before a gas company will purchase it.

Oil treating

Crude oil is produced with various quantities of gas, water, and other impurities mixed with the oil. Each of these impurities must be separated out before the oil can be sold. This process is known as oil treating, and oil treating systems are important parts of lease equipment.

The kind of treatment system chosen depends on several factors:

- Tightness of emulsion.
- Specific gravity of the oil and produced water.
- Corrosiveness of the oil, gas, and produced water.
- Scaling tendencies of the produced water.
- Quantity of fluid that is to be treated and of water in the fluid.
- Availability of a sales line to sell the gas.
- Desirable operating pressure for the equipment.
- Paraffin-forming tendencies of the crude oil.

Emulsions are mixtures of fluids. Usually they are water in oil; however, some are oil in water, or "reverse emulsions." To break a crude oil emulsion to separate the clean oil, you must displace the emulsifier and its film. This allows the water to coalesce into heavier droplets so they can drop out of the oil.

Water treaters

One of the most common types of water treaters is the heater treater (Fig. 11-4). Heater treaters use thermal, gravity, mechanical, and sometimes chemical and/or electrical methods to break emulsions. Heater treaters can be vertical or horizontal, and their size depends on the volume of water and oil to be handled. Their main function is to heat the water, which helps break up the emulsion.

Other kinds of treaters are those equipped with electrodes, sometimes called electrostatic coalescers or chem-electric treaters. These kinds of treaters are good because they treat at a lower temperature, saving fuel and oil gravity.

Free-water knockout (FWKO)

This vessel is used to separate free gas and free water from free oil and emulsions (Fig. 11-5). Its size depends on the desired retention time and the volume of water per day to be handled. Time, gravity, mechanical, and sometimes chemical methods are used to hasten separation when FWKOs are used.

When heat must be used to break an emulsion, significant fuel gas can be saved by using the FWKO. Heating water unnecessarily is useless as well as costly.

Fig. 11-4
Flow diagram for a vertical heater treater.

Gun barrel

Sometimes an oil-water emulsion is not stable. Given enough time, the water will settle to the bottom of a tank and the oil will rise to the top. The settling vessel used for this kind of separation method is called a *gun barrel* or *wash tank* (Fig. 11-6). Although there are various designs of gun barrels, usually they must be tall enough to allow the clean oil to gravity-flow into the stock tanks. The water is drawn off through the water leg at the bottom of the tank.

STORAGE TANKS

Once the oil is clean enough to meet pipeline specifications, it is flowed into storage tanks, sometimes called *stock tanks*. This entire configuration of

Fig. 11-5
Free water knockout with oil split option.

separating equipment, treating equipment, and storage facilities is called the *tank battery*.

There are two basic types of stock tanks: bolted steel and welded steel. Bolted steel tanks are usually 500 bbl or larger and are assembled on location. Welded steel stock tanks range in size from 90 bbl to several thousand bbl and are welded in a shop and then transported as a complete unit to the site. Welded tanks can be internally coated to protect them from corrosion, and bolted tanks offer the option of internal lining or galvanized construction for corrosion protection.

Vapor recovery system

When oil is treated under pressure and then goes to a stock tank at near atmospheric pressure, some of the liquid hydrocarbons flash, or convert, to gas. In past years, flash gas or vapors vented to the atmosphere. Government agencies now insist on vapor recovery in order to reduce air pollution.

A vapor recovery unit consists of a control pilot mounted on a tank for compressor control, a scrubber to keep the liquid hydrocarbons out of the compressor, a compressor, and a control panel. Only one ounce of gas pressure is needed to start the electric motor-driven compressor, and it shuts off automatically at about 1/4 ounce gas pressure.

GAUGING AND METERING PRODUCTION

A lease operator must measure the proper amount of oil or gas from the wells on a lease to see that proper credit is given for oil and gas delivered from

Fig. 11-6
Schematic diagram of wash tank or gun barrel.

the lease. For the best control, volumes of oil, gas, and salt water produced are usually checked or measured by the lease operator or gauger during a 24-hr period. When a tankful of oil is delivered or "run" to a pipeline, tank car, or tank truck, the oil is measured by *gauging* the height of oil in the stock tank before and after delivery. The oil is tested to determine its gravity because the value of crude varies with the gravity specification. The temperature of the oil and its BS&W content of sediment and water also are determined.

Sometimes an oil meter is used rather than a tank for testing the well. This way, oil production from all wells on a lease can be measured in the lease tank while one well is individually tested using an oil meter or test tank. State regulatory agencies require periodic tests. The operator uses them as a guide to maintain efficient operation of the wells and the underground reservoir.

Tank strapping

Before a tank battery is put into service, the storage tanks are strapped. This means the tanks are measured, and the volume of oil that can be contained in each interval of tank is computed. The capacity in barrels, according to the height of the liquid in the tank, is prepared in tabular form, known as a *tank table.*

Gauging

When gas passes through the oil-gas separator, it is measured through an orifice meter. The crude oil flows a separate direction into one of the tanks where the operator measures the height of oil. To measure or gauge the level of oil in a tank, the gauger lowers a steel tape with a plumb bob on the end into the tank until it touches the tank bottom. The tape is then withdrawn. The highest point where oil wets the tape shows the level or height of oil in the tank. When this value is compared to the tank table, the gauger can determine the volume of oil (or oil and water) in barrels.

Another gauging device is an automatic tank gauge. This device is a steel gauge line contained in a housing with a float on the end of the line, resting in the surface of the oil in the tank. The line extends over the top of the tank and down the outside through a *reading box*. The line is marked to show the height of the oil in the tank, and it can be read through a glass window in the box.

Final gauging, or top gauging, the stock tank is made by a pipeline gauger before running the oil from the tank into the pipeline. The lease operator is responsible for watching the gauging and testing of the oil by the pipeline gauger to be certain the measurements are accurate. The pipeline measures the volume of crude, the BS&W content, the temperature, and the API gravity. These measurements are written up on the pipeline *run ticket*, which is signed by both parties.

Lease automatic custody transfer (LACT) units

LACT units, as they are commonly called, are automatic measuring devices where oil is measured for sale and transfer to the pipeline. This eliminates much of the pipeline gauger's job; however, the gauger must calibrate the units periodically and check on maintenance.

In addition to LACT units, other electronically controlled devices are widely used to control and monitor production. These include automatic well test systems, fail-safe devices, sensors, time clocks, and alarm devices. When properly checked and maintained, automatic systems can increase the productivity of surface equipment systems.

Fig. 11-7
General lease flow diagram.

CHAPTER 12

Production Problems and Workover Operations

Of all the problems that can occur during production, three stand out the most: equipment failure, wellbore problems, and salt water disposal. Occasionally, these problems result in a workover operation. Let's look at these common production foul-ups and learn more about them.

EQUIPMENT FAILURE

Equipment failure is probably the most common type of production problem. For example, a rod may break in a pumping well, requiring special equipment called a service or *pulling unit* to be moved to the well to recover the rod from the hole and put the well back on production (Fig. 12-1). The pulling unit is usually mounted on a truck—or, if it is a very large unit for deep wells, on a trailer—and has a special crew of its own. If the well does not have a derrick, and most wells drilled in recent years do not, the unit will include a mast and winch for removing equipment from the wellbore.

Another common production problem is subsurface pump failure, due in most cases to physical wear of one or more of the pump's moving parts. When this occurs, a pulling unit can quickly remove the pump, which is attached to the sucker rods, and make the necessary repairs.

If tubing develops a leak or a break from corrosion (Fig. 12-2) or mechanical stresses, the pulling unit is again called to the wells. The tubing is removed from the hole, the damaged joint is replaced, and the tubing is returned to the wellbore.

In gas lift operations, mechanical failure of gas lift valves is common. A valve may become stuck in either the open or closed position; in either case, it

Fig. 12-1
A service or pulling unit, which recovers broken rods from downhole.

Fig. 12-2
A section of corroded tubing. Corrosion is one of the more costly production problems.

must be promptly removed and repaired. One type of gas lift valve is installed by a wireline in a specially designed pocket in the tubing known as a *gas lift mandrel*. When a failure occurs in this type of valve, it is not necessary to remove the tubing. Instead, a small truck equipped with a winch and a wireline retrieves and replaces the faulty valve (When a failure occurs in the conventional type of gas lift valve, the entire tubing string must be removed in order to replace the damaged equipment.)

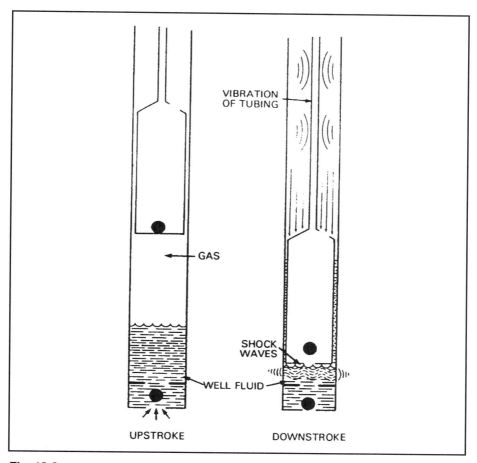

Fig. 12-3
Fluid pounding.

Fluid pounding

When a well being pumped has been on production long enough to stabilize, the ideal situation occurs when the inflow rate of the well matches the
pumping rate with sufficient pump submergence to permit the pump to be
completely filled with each stroke. The pump can be completely filled, or nearly so, only when the gas being produced is separated from the well fluid and
vented through the casing when the pumping rate and inflow rate are the
same. A gas anchor below the pump assists in the bottomhole gas separation
(Fig. 12-3). In general, casing pressure should be maintained as low as possible

to create a pressure sink in the bottom of the well as compared with the formation pressure. If the pump rate exceeds the inflow rate, the well will pump off, and the pump will not fill completely on the upstroke. On the downstroke, the pump plunger will strike the incompressible fluid or liquid, causing fluid pounding. The result is shock loading on the rod string, the surface equipment, and the gears.

Severe fluid pounding is easily detected from the dynamometer card or by vibration of the polished rod. It should be corrected by reducing the number of strokes per minute or by reducing the stroke length. If the well continues to pump off after the pumping unit's lower limit has been reached by reducing speed and reducing stroke length, fluid pounding can be reduced or eliminated by pumping the well intermittently. The unit can be turned off and on manually, with a time-cycle controller, or with a pump-off controller. The result of fluid pounding can be costly repairs and downtime. Intermittent pumping can reduce these costs. However, a pumping system properly sized for the inflow of the well will be more productive and cost effective.

WELLBORE PROBLEMS

Sanding, formation damage, paraffin accumulation, oil-water emulsions, and corrosion are common wellbore problems.

Sanding

In wells that produce from loosely consolidated sandstone formations, a certain amount of sand is usually produced with oil. Although some of this sand will be produced at the surface, most of it will accumulate at the bottom of the hole. Continued accumulation of the sand in the wellbore will eventually cut the oil-producing rate and may even halt production altogether. When this problem, known as sanding, occurs, a pulling unit equipped with a sand pump on a wireline is called to the scene. The sand pump is a special type of bailer that removes the sand from the wellbore.

If a well continues to present sanding problems, preventive action may be needed. One of the most frequently used methods of combating sanding is to gravel-pack the well. When a well is gravel-packed, a slotted liner is installed opposite the producing interval and carefully graded gravel is placed outside and around the liner. The gravel is larger than the sand grains of the formation but small enough that these sand grains cannot move through the gravel bed. Thus, the gravel forms a bridge through which the oil—but not the sand—can flow.

Various types of plastics are also used to consolidate or compact the sand. The chief problem here is to obtain a plastic that will consolidate the sand yet permit oil to flow through the result.

Formation damage

This common problem occurs when something happens to the formation near the wellbore, slowing oil production. For example, excessive buildup of water saturation in the vicinity of the wellbore impedes oil flow. A mud block, an accumulation of drilling mud around the wellbore producing zone, can also reduce the rate of oil flow. In a shaly producing formation, the drilling mud used in a workover operation can cause clay swelling and completely stop oil flow.

Wells with such formation damage may be treated with acids, mud cleanout agents, wetting agents, and/or other special-purpose chemicals. These materials are pumped into the formation and are eventually produced to the surface. These are highly specialized operations, requiring special pump trucks and equipment, and they are usually performed by oil well service companies specializing in this type of work.

Paraffin

Paraffin accumulation in the tubing and the surface flow lines is a problem in some areas in which a special type of crude oil known as "paraffinic crude" is produced. Paraffin, which is actually a part of this crude oil, precipitates in solid form as a result of temperature reduction. Thus, paraffin accumulation is seldom a problem at the bottom of the hole, but it becomes acute near the surface where the temperature is lower.

Various methods are used to deal with this problem. In surface flow lines, it may be sufficient to pump scrapers through the lines at periodic intervals to remove the accumulated paraffin. In tubing, scrapers can be installed on the sucker rods, whose up-and-down motion will keep the scrapers moving and thus keep the tubing free of excessive paraffin accumulation.

Another way to remove paraffin is to circulate hot oil through both the surface lines and the tubing at periodic intervals. Such hot-oil circulation is usually performed by a service company, since this is another specialized operation that is required only from time to time.

A chemical called paraffin solvent can also be injected down the casing-tubing annulus to prevent the accumulation of paraffin.

Oil-water emulsions

Emulsions of oil and water are a fourth common production problem. Under certain conditions, oil and water may form an emulsion that will not separate at the surface without special treatment. This is a problem because the process to break up the emulsion is very expensive. Methods of breaking up such emulsions include heat treatment, chemical treatment, and various combinations of chemical treatment. Since the chemical composition of crude oil varies from one field to another, the nature of the chemicals used to break up emulsions also varies.

Corrosion

Corrosion of equipment is one of the most costly problems plaguing the oil industry (Fig. 12-2). Salt water produced with oil is highly corrosive, and most crude oils contain varying amounts of hydrogen sulfide, which is also quite corrosive. Anticorrosive measures include the injection of a chemical corrosion inhibitor down the casing/tubing annulus; the use of plastic-coated tubing; and the use of special corrosion-resistant alloys and cement-lined pipe. Each of these methods has distinct advantages and disadvantages. Frequently the cost of reducing the corrosion rate is so high that it cannot be justified, in which case no anticorrosion measures of any kind are taken, and the equipment is replaced at the end of its useful life.

SALT WATER DISPOSAL

Disposing of the salt water produced with oil can be very expensive. Salt water cannot be run into surface streams and pools because it is harmful to plant and animal life. The most common method of salt water disposal is through wells drilled especially for this purpose.

Salt water must not be injected into a freshwater formation, and wherever injected it must be handled with care to prevent the accumulation of excessive foreign materials which might plug the formation. It is common practice to back-flow salt water injection wells from time to time to remove some of the foreign materials that have accumulated on the formation at the bottom of the hole. Acidizing the injection well also helps to clean the formation.

WORKOVER OPERATIONS

Workover operations are major remedial operations sometimes required to maintain maximum oil producing rates. If, for example, a well begins to produce an excessive amount of salt water, a workover rig—very similar to a drilling rig but somewhat smaller—is moved onto the well, and operations to reduce the saltwater production are begun.

It is first necessary to "kill" the well with some fluid, such as drilling mud, salt water, oil, or possibly a special workover fluid, which has sufficient hydrostatic pressure to counteract the formation pressure when the hole is filled with the fluid. If the salt water is coming from the lower part of the reservoir, it is usual to squeeze-cement the perforation with either a low-pressure or a high-pressure squeeze.

If the high-pressure squeeze-cementing technique is used, a special packer is run on the bottom of the tubing to protect the casing and other equipment at the wellhead. If the low-pressure or "bradenhead" squeeze-cementing method is used, then a packer is not required, since the pressures applied will not exceed the working pressure of the wellhead equipment and casing. After the cement has set, it may be necessary to drill out the cement from inside the casing and reperforate the casing at the desired intervals, since the cement will have sealed off all the old perforations.

If a well is producing with an excessive gas-oil ratio, it may be possible to reduce the gas-oil ratio by the same squeeze-cementing and reperforating technique.

Where there is more than one producing interval in the wellbore and a lower zone has been depleted, a plugback to a high zone is in order. The *plugback* can be accomplished with a cement plug in the casing or with a bridge plug—a mechanical device which can be set in the casing to effectively seal off all production below the point at which it is set.

The so-called permanent completion permits all workover operations to be conducted with wireline equipment, eliminating the need for workover rigs. Permanent completion equipment features special types of valves that can be opened and closed by wireline equipment. A complete line of equipment has been designed for this type of workover operation, and even cementing and reperforating can be satisfactorily accomplished with it.

Either during workover or during initial completion, a well may need to be stimulated to increase production rate. That's the topic of the next chapter.

CHAPTER 13

Stimulation Methods

When tests indicate that a well may be an economical producer, but for some reason the rate of flow is inadequate, the formation may be stimulated to increase the well's productivity. The earliest type of stimulation method used was shooting with nitroglycerin. Nitro was lowered in the wellbore, and the resulting explosion created cracks and fissures in the formation. Production was generally improved, but the wellbore was destroyed.

Today, nitro shooting is somewhat more sophisticated, but countless new stimulation methods have come to the forefront. Among the most popular of these methods are acidizing and formation fracturing.

ACIDIZING

Acid was first used for well stimulation in 1895. Acid, pumped into the microscopic flow channels of the rock formations, will dissolve the rock and enlarge the passages. This improves the flow of reservoir fluids to the wellbore. Although significant production increases were obtained, acid solutions proved extremely corrosive to well equipment, so the method was abandoned.

The development of chemical inhibitors that allowed acid solutions to react selectively with formation rocks without attacking well equipment revived interest in oil well acidizing in 1932. The excellent results obtained from the improved acid stimulation process increased the use of the procedure, which is one of the standard completion and remedial techniques today.

Hydrochloric acid is most commonly used for acidizing treatments because it is economical and leaves no insoluble reaction product. Hydrochloric acid by weight contains about 32% hydrogen chlorine gas. The acid is kept in storage tanks and is diluted to the desired concentration—usually about 15%—prior to treatment.

When hydrochloric acid is pumped into a lime formation, a chemical reaction takes place. The rate of reaction during an acidizing treatment is proportionate to the acid concentration and temperatures, and inversely proportionate to the pressure. But because considerable pressure is required to return highly viscous spent-acid solutions from the pores of the formation, concentrations greater than 15% are rarely used in acid treatments.

Acid strength can be estimated in the field by either a hydrometer or a field titration kit. The accuracy of the hydrometer readings depends on the care and technique used by the engineer. When testing, the hydrometer and the glass cylinder must be cleaned carefully to ensure that no dirt or oil remains on the moving parts. The temperature of the acid sample should be corrected to 60° F.

Preliminary testing

When acidizing, several characteristics must be evaluated; that is why testing is so important. Cores or cuttings provide information on porosity, permeability, and oil and water saturations. A sample of the crude from the formation can also be tested for emulsifying tendencies. If the crude tends to emulsify with either the fresh acid or the spend acid, appropriate emulsion-breaking additives will be required.

Another important factor is to determine the swelling properties of silicate components of the formation rock. In some cases, clay and bentonite particles may enlarge to several times their original volume after being exposed to an acid solution. These enlarged particles may plug the microscopic flow channels in the reservoir or, even worse, may reduce the flow channel to smaller than original. So if tests indicate the formation sample exhibits a tendency to swell, proper silicate control additives are used to prevent this swelling and the resulting damage.

Acidizing equipment

Specialized transport and pumping equipment has been developed for acidizing oil and gas wells. Tank trucks, ranging in size from 500 to 3,500 gal capacity, carry the acid solutions to the well site. Chemical additives are premixed into the acid at the time the transport is loaded.

Truck-mounted pumps are used to inject the acid down the well and into the productive formation (Fig. 13-1). The pumps' heavy-duty gasoline or diesel engines are capable of producing as much as 1,000 hydraulic horsepower. These large horsepower ratings are necessary to force the acid into the pores of the rock against the natural formation pressure.

Treating techniques

The two basic types of acidizing treatments are the uncontrolled, or nonselective, method and the controlled, or selective, method (Fig. 13-2).

Fig. 13-1
An acid pumping trailer (courtesy Halliburton).

In *uncontrolled treatment,* an acid solution is pumped down the casing and is followed by enough displacement fluid to force the acid out into the formation. This method can be done with or without tubing in the well and is most applicable in single-producing-zone wells, water injection or disposal wells, low-pressure gas wells, or low-capability wells. The advantages are savings in costs and time. Also, the reaction products can be removed more readily from the producing formation. The disadvantage is that there is no control over where the acid will go. Stimulation fluid could be wasted on an unproductive zone.

The treating procedure is as follows:

- Remove the fluid in the hole by either swabbing or bailing.
- Pump the acid into the casing. If the fluid has not been removed, it will be forced into the formation ahead of the acid.
- Follow the acid with sufficient displacement fluid to force all the acid out into the formation. The pressure used to force the acid into the formationis limited by the size and capacity of the surface pumps.
- After sufficient time has elapsed for the acid to react completely, remove the spent acid containing the reaction products from the well by swabbing,bailing, pumping, or, if there is sufficient bottomhole pressure, by flowing the well.

In the case of water injection wells, it is often possible merely to resume injection so the spent acid is forced out into the formation away from the wellbore. There it cannot interfere with subsequent operations.

In the conventional *controlled acidizing treatment*, tubing must be in the hole and the well must be capable of being filled with fluid. The tubing is positioned below the productive section. The well is first filled with oil, then is followed by enough acid to displace the oil in the tubing, including the annular volume over the pay section. As soon as the acid is positioned opposite the *pay*, or *producing*, section, the casing outlet is closed. The acid is pumped down the tubing and is forced out into the formation. The acid is then followed by enough displacement fluid to clear the tubing and wellbore.

Another type of controlled treatment is the *packer method* (Fig. 13-3). Here, a packer is run in on tubing to a joint just above the zone to be acidized. The

Fig. 13-2
An uncontrolled (a) and a controlled (b) acidizing method.

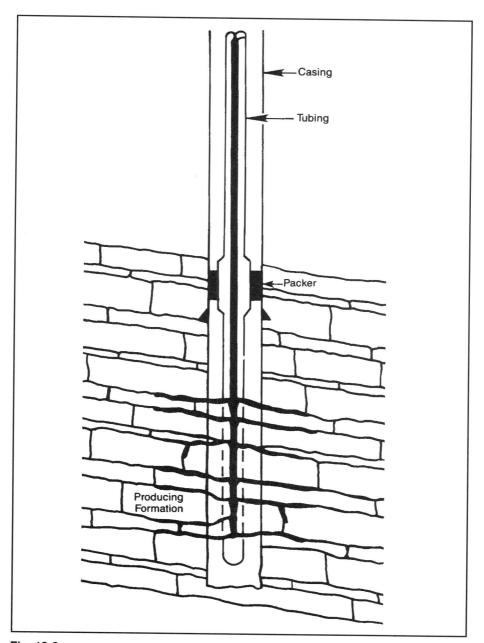

Fig. 13-3
Acidizing a well using the packer method.

well is filled with oil, after which acid is pumped down the tubing and positioned across the productive interval. The packer is then set, preventing the acid from traveling further up the annulus.

Sometimes the packer is set first and the oil is swabbed from the tubing. After this, the acid is pumped down. In some cases, the acid is pumped into the tubing, displacing the oil ahead of it into the formation.

The advantage of the packer method is that the acid is confined to the section of the formation below the packer. This prevents entry into the nonproductive zones up the wellbore. If necessary, oil may be pumped into the annulus to reduce the pressure differential across the packer and prevent it from becoming unseated.

Other popular forms of controlled treatments include the selective electrode technique, the radioactive-tracer technique, the combination method, ball sealers, and temporary plugging agents. All of these methods have their own advantages and disadvantages and should be carefully analyzed before use.

Overall, the advantage of selective acidizing is that maximum benefit is derived from the acid by controlling the section into which it enters. In addition to keeping the acid out of unproductive zones, the acid can be diverted to the less-permeable sections that otherwise would not benefit from the acid. Also, the acid can be diverted away from any known water zones that would not benefit from the acid.

The disadvantages of selective acidizing are that it costs more, the treatment is harder to conduct, and (in some cases) more time is required for the well to clean up after treatment.

Stage acidizing

Stage acidizing is for treating dense and tight limestones. The well is treated with two or more separate stages of acid rather than one large treatment. This allows the work to be done at lower pressures than would be necessary with a single large treatment. Generally the acid is swabbed out of the well between stages to avoid pushing the spent acid farther into the formation.

Sometimes in limestone formations, when there is a possibility of breaking into a water zone, stage treatments are used. This permits the treatment to be stopped at the first sign of water. The spent acid is checked for water after each stage of acid.

Another application is to clean up the contaminated zone near the wellbore with one stage of acid. Then later stages can readily penetrate farther into the formation at lower pressures.

If a formation contains fine insoluble sand or chert particles that could cause plugging, sudden pressure increases are often encountered during conventional treatments. When plugging occurs, the acid must be circulated out of the tubing and the well cleaned up before the treatment can proceed. Stage treating alleviates this problem because the fresh acid in each succeeding stage can penetrate the formation at lower pressures and higher rates.

ACIDIZING ADDITIVES

The physical and chemical characteristics of the formation rock often affect the results of an acid stimulation treatment. In some cases, special additives improve the action of the acid or avoid clean-up difficulties in recovering the spent, or old, acid following the job.

Inhibitors

Inhibitors are dissolved in acid solutions to slow the reaction rate of the acid with metals. They are necessary to avoid damage to casing, tubing, pumps, valves, and other equipment. Inhibitors do not completely stop the reaction between the acid and the metal, but they reduce the metal loss from 95-98%. These reactors do not affect the reaction rate with limestone, dolomite, or acid-soluble scales. All acid used in such treatments today are mixed with one of these inhibitors. Inhibitors come in two types. One type is an organic inhibitor, such as nitrogen- or sulfur-bearing organic compounds. A second type is an inorganic, primarily copper. Arsenic was used in the past but has been discontinued.

Intensifiers

Intensified acid is a mixture of inhibited hydrochloric and hydrofluoric acids. The fluoride speeds up the reaction rate of the acid and enables the acid to dissolve otherwise insoluble minerals found in dolomite.

Intercrystalline films of silica, insoluble in hydrochloric acid, often exist in the crystalline structures of dolomite. When present, they prevent the acid from contacting the soluble portions of the rock. Hydrofluoric acid dissolves the silica, allowing the hydrochloric acid to reach the soluble portions of the rock.

Surfactants

Surfactants are chemical additives that lower the surface tension of a solution. The acidizing solution's efficiency is enhanced by adding the desired surfactants.

Introducing a surfactant helps the acid solution penetrate the microscopic pores of the rock. The acid's improved penetrating ability results in deeper penetration of the formation and improved drainage following the treatment.

In addition, surfactants let the acid penetrate oil films surrounding the rock and lining the pores, permitting the acid to contact the rock to dissolve it.

Surfactants also facilitate the return of spent acid following the treatment. It is important that none of the acid remains behind to block flow channels. A surfactant in the acid allows a more complete wetting of the rock; it also reduces resistance to flow of the acid. The spent acid is readily returned through the treating section. This procedure is especially important in low-pressure wells.

A secondary advantage of surfactants is that a demulsifying action is obtained. The surfactants inhibit the occurrence of emulsions or destroy those already formed.

Surfactants in acid solutions have resulted in the removal of considerable amounts of infiltrated brine with the spent acid. This leaves the formation free from contaminants that might have restricted the production capability of the well.

Demulsifiers

Many of the components naturally occurring in crude oils possess emulsion-forming and stabilizing properties. When crude oil is mixed or agitated with acid or spent acid, emulsions may be formed. In some cases these block the formation, reducing or even preventing production from the well. Demulsifiers, when added to an acid solution, are chemical agents that counteract natural emulsifiers in crude oil.

Silicate control

Silicate components—clays and silts—are present in most limestones and dolomites. One characteristic of silicates is that they swell in spent acid. Naturally this reaction is undesirable. Swollen particles may block the formation flow channels and reduce production rates.

Silicate-control additives are chemicals designed to prevent freed silicate particles from absorbing water. Some chemicals prevent acid solutions from spending beyond the pH at which silicate particles occupy the smallest possible volume. Other chemical additives shrink silicate particles by replacing that which is absorbed with a water-repellent organic film. The engineer can control formation plugging, lower treating pressures, accelerate cleanup times, and reduce the occurrence of particle-stabilized emulsions by using the proper chemical additives for silicate control.

Hot acid

Hot acid solutions are beneficial in wells where the formation or scale deposits within the wellbore are slow to dissolve and difficult to remove. By heating the acid, the reaction time is decreased and a more effective treatment is achieved. This treatment is particularly valuable in wells where mineral deposits on screens and well equipment interferes with production. It is also effective for enhanced recovery injection wells that are partially plugged with difficult-to-dissolve minerals.

Sometimes organic solvents and oils are used in conjunction with hot acid treatments. A combination of heat and solvent action is effective when considerable deposits have accumulated in formation flow channels, blocking production.

First, the gelled solvent oil, carrying suspended magnesium pellets, is injected into the pay zone. A conventional hydrochloric acid solution plus any necessary additives is then pumped into position. After the acid and magnesium react, the formation temperature may increase to 200-300° F. The formation is usually cleaned quickly by the combined action of the acid on mineral deposits and the heat and solvent action on paraffin, asphalt, and tar deposits. In addition, the hydrogen from the magnesium and acid reaction produces turbulence that dislodges particles wedged in flow channels, also aiding the cleaning process.

Retarded acid

In some highly reactive rock formations, the acid's reaction time is slowed to increase penetration instead of spending most of the acid in the immediate vicinity of the wellbore. Many different gums, thickening agents, and other inhibitors slow the reaction time of the acid and permit deeper penetration.

Some retarded acids contain chemicals that deposit a film over the rock following the initial reaction between the acid and the formation. In other cases, the high viscosity of the thickened acid produces the desired result. Acid-oil emulsions of controlled stability (to assure breakout after a predetermined time) have also been used to obtain retarded acidizing action. In many cases, the acid is prevented from entering the smaller pores and, under pressure, enters only the largest pores. As a result, the surface area in contact with the acid is restricted and deeper penetration is obtained before the acid is completely spent.

Often a quantity of retarded acid is used to produce channels radiating from the wellbore. This treatment is followed by additional hydrochloric acid to enlarge the newly formed flow channels. The advantages of using retarded acid are that the well's immediate drainage area is greatly increased and maximum benefit is derived from the acid during treatment. Also, less formation pressure is required to clean up reaction products following the treatments.

Iron retention

Water injection wells for secondary and enhanced recovery projects (see Chapters 14 and 15) or for brine disposal purposes frequently are subject to plugging formation flow channels. Usually the condition can be relieved by hydrochloric acid. However, the iron compounds that are dissolved will reprecipitate as a bulky, gelatinous hydroxide when the acid is spent. Unless countermeasures are taken, serious plugging may occur.

Chemicals called *complexing agents* chemically tie the dissolved iron into complex ions. In most cases, reprecipitation of dissolved iron as a hydroxide is eliminated completely.

Mud removal acid

Mud used to remove drilling fluid is a mixture of hydrochloric and hydrofluoric acids, containing required inhibitors, surfactants, and demulsifiers. This acid is called mud acid, and it dissolves clays commonly used in drilling fluid.

Mud acid removes the mud cake from the face of the productive interval during completion and before subsequent workover jobs. It can also remove infiltrated drilling mud that may block formation flow passages. The acid disintegrates the accumulation of mud deposits, leaving the face of the productive interval free and clean. Mud acid also increases the permeability of sand formations. If laboratory tests indicate the solubility of the rock in mud acid is higher than other acids, then treatment is suggested.

A mud acid treatment may be preceded by a wash acid containing 15% hydrochloric acid with inhibitor, surfactant, and demulsifier. The procedure removes any readily soluble material from the face of the productive interval. This ensures that the mud acid treatment will react with the difficult soluble portions of the rock.

Cleaning solutions

Frequently cleaning solutions are used before fracturing, cementing, and acidizing treatments. The cleaning operation provides uniform distribution of the stimulation treatment over the entire vertical extent of the productive interval. Cleaning solutions are acid mixtures that contain no fluorides.

Dry acid

Dry acetic acid employs an oil-soluble nonaqueous acid for acidizing treatments. Acetic acid is blended into oil and is injected into the rock formation in the same manner as other acids. It will not react with the rock until water is encountered in the formation. The small amount of connate water in the rock pores allows the acetic acid to react with carbonates in the formation.

Table 13-1
Chemical Stimulation Check Sheet

TYPE OF ACID	APPLICATION	SYMPTOM OR TYPICAL PROBLEM	PROPERTIES	BENEFITS	OTHER CONSIDERATIONS
Regular HCl-Inhibited	Oil, gas, water or injection wells in all types of limestone and dolomite formations.	Low production, low effective permeability with production decline or low injectivity.	Will dissolve 10.0 cu ft limestone or dolomite per 1000 gal 15% HCl.	Helps increase productivity in limestone and dolomite types of formations.	Concentrations 5 to 30%. Compatible with additives.
Penetrating or Pen-Acid	Gas, oil, and injection wells.	Low injection, expected high treating pressure. Carbonate scale buildup. Oil contamination.	Surfactant blended with regular HCl for better penetration. Lower surface tension.	Faster cleanup. Lower treating pressures improved injectivity. Economical.	Used ahead of fracturing. Used with Gypsol. Use FE Acid for iron scales. For improved foaming use Howco-Suds.
Non-Emulsifying or N.E. Acid	Oil or gas condensate wells.	Low production. High basic sediment in produced fluid. Existing or potential emulsion blocks.	Regular HCl plus non-emulsifying and low surface and interfacial tension properties.	Faster cleanup by breaking or preventing emulsions.	Used in calcareous formations producing oil or distillates.
HV-60 Retarded Acid	Fracture acidizing in moderate to high temperature wells.	Low effective permeability and lack of drainage area.	Oil external emulsions. Retarded HCl. High viscosity.	Delayed chemical reaction. Deeper penetration.	Compatible with temporary blocking agents. Good viscous preflush. Reaction time can be varied.

201

Table 13-1
Chemical Stimulation Check Sheet

TYPE OF ACID	APPLICATION	SYMPTOM OR TYPICAL PROBLEM	PROPERTIES	BENEFITS	OTHER CONSIDERATIONS
TGA (True Gelled Acid) Retarded Acid	Low temperature wells.	Low effective permeability. Low productivity and need for increasing drainage area.	Gelled regular acid. Moderate viscosity.	Delayed chemical reaction.	Low friction and low fluid loss properties.
CRA (Chemically Retarded Acid)	All types of wells. Matrix or fracture acidizing.	Need for low pressure acidizing with extended reaction time to stimulate extensive drainage area.	Retardation without viscosity increase.	Delayed chemical reaction. Low viscosity. Can be modified to suit treating requirements.	Compatible with other acid additives. For all special purposes requiring a slow acting retarded acid. Provides deeper penetration.
MOD 101, 202, 303 Retarded Acid	All types of wells.	Need to increase production. Well may have been unsuccessfully treated with other acid.	Unusual etching. Slow reaction. Lower corrosion. Less sludging.	Provides deeper penetration in fractures or true permeability.	Compatible with most other acid additives. Can be further retarded with CRA. Holds iron in solution.
HF Acid (HCl-HF)	Damaged sandstone formations.	Pressure buildup tests show damage, low productivity or drop in productivity.	Dissolves clay minerals and silica.	Can remove shallow damage caused by mud, etc.	Must be used with proper pre and after flush and compatible surfactants.
Orangic HF (Acetic-HF) (Formic-HF)	Damaged sandstone formations. Higher bottom hole temperature.	Damage from mud or fluid lost to formation.	Dissolve clays and sand.	Helps remove deeper damage because of retardation.	Also effective in removing (shallow) mud damage at higher bottom hole temperatures.
SGMA* (Self Generating Mud Acid)	Remove deeper damage.	Low productivity. Damage from movement of fines or fluid invasion.	Retarded HF Acid.	More retarded than organic HF.	Usually shows no emulsion problems.

Product	Application	Problem/Condition	Properties	Action	Notes
CLAYSOL	Removes deeper damage.	Fines movement or clay swelling causes low productivity.	Retarded clay solvent.	Dissolves clay in preference to sand.	No limitation on depth of damage which can be removed. No temperature limitation.
MCA (Mud Cleanout Agent)	Stimulation of all types of wells. Ahead of cement.	Low apparent permeability from mud filtrate, mud solids, water or emulsion invasion.	Low surface and interfacial tension, emulsion breaking and solids dispersing properties.	Disperses and removes whole mud and filter cake. Breaks emulsion and water blocks.	Broad spectrum application, works with most oils and muds. Can be used ahead of cement to improve bond. Cleans up perforations.
MSA (Multiple Service Acid)	Stimulation of all types of wells.	Scale buildup in pumping wells. Low productivity in high temperature wells. Need low corrosive perforating fluid.	Slow reacting organic acid. Low corrosion. Will not strip chrome plate.	Can be dumped down annulus of pumping well. Used as perforating fluid or as retarded acid in higher temperature calcareous well.	Does not cause hydrogen embrittlement. Minimal damage to wireline and perforating equipment.
OSA (Oil Soluble Acid)	Where aqueous fluid may damage formation.	Clay swelling, scale, water or emulsion blockage.	Organic acid in organic solvent. Utilizes interstitial water.	Provides acidic environment without introducing additional water.	Also removes oil base mud or organic deposits with minimal corrosion.
FE Acid	Oil, gas, flood or disposal wells.	Production impaired by corrosion products, iron scales or secondary reaction products.	Sequesters iron and controls pH.	Helps prevent secondary deposition of iron reaction products.	Can be modified with most other acid additives. Low pH minimizes clay swelling.
PAD (Paragon Acid Dispersion)	Producing or water injection wells. Converting old producers to injector.	Deposit build up composed of scale, paraffin, asphaltene clays, and/or corrosion products.	Aromatic solvent dispersed in acid.	Removes organic and inorganic deposits simultaneously.	Type acid can be varied to correspond to deposit.

FORMATION FRACTURING

In formation fracturing, developed about 1948, oil or water, mixed with sand or another propping material, is pumped into the formation at a high rate, causing fractures in the formation. The sand moving with the water through these fractures actually props them open. This significantly increases the drainage radius of the wellbore and thus the productivity of the well.

Fracturing has been successful in all types of formations except those that are very soft and unconsolidated. The plastic nature of soft shales and clays makes them difficult to fracture.

The production increase from a fracture treatment varies considerably, although it usually averages about 200-300%. A much greater increase may be obtained if permeability blocks around the wellbore restrict production. As far as ultimate recovery, the technique can increase recovery by 5-15%. Thus, fracturing makes production profitable from many wells and fields that could not have been profitable otherwise.

Fractures and fracture patterns

A fracture occurs in the wellbore when hydraulic pressure overcomes the combined resistances of the formation's tensile strength and the compressional stresses caused by the weight of the overburden. A fracture begins at the point where the sum of these two forces is the least. In shallower formations, a horizontal fracture is usually produced; in deeper formations, a vertical fracture is produced (Fig. 13-4).

Fractures formed during a fracturing, or frac, treatment must open wide enough to accommodate the flow or proppant-laden fracturing fluid. The walls of the fracture tend to close after the treatment, so sand and some other proppant must be left in the fracture to hold it open.

Fracturing equipment

The four basic types of fracturing equipment are pumpers, blenders, sand transports, and fluid transports (Fig. 13-5). In the early days, the equipment could only pump 40 gal/min at 5,000 psi. Today, advanced units can operate for extended periods at pressures as high as 20,000 psi, and many units can be combined for a single job.

The surface pressure required to fracture a well depends on a combination of three factors:

- The pressure needed to inject a fracturing fluid into the formation at the bottom of the well.
- The friction pressure losses encountered from the flow of fracturing fluid down the tubing or casing.

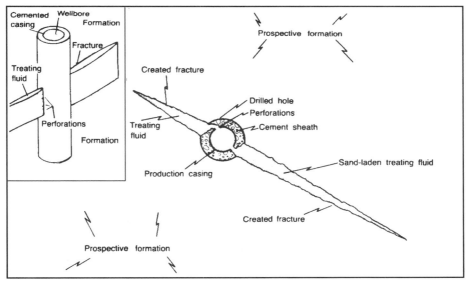

Fig. 13-4
Concept of hydraulic fracturing.

• The pressure exerted by the column of fluid in the hole.

The total surface pressure required equals the formation pressure plus the pressure drop in the pipe due to friction less the fluid head of the fracturing fluid. In many cases, especially when fracturing through tubing, friction is the most important consideration.

Fracturing techniques

Most early treatments were made through tubing and below a packer. The method is still used today when excessively high pressures are expected or when the casing may not withstand treating pressures. However, with the trend toward higher injection rates and larger fracturing jobs, friction pressures in the tubing become very high and limit rates. Sometimes this causes *screen-outs*, when the sand begins to fall out of the liquid and fills up the bottom of the well.

To overcome large friction losses, the tubing is removed and the treatments are performed down the casing. Injection through the casing allows increased injection rates. Another practice is to treat down the annular space at the same time. This allows the tubing to be left in the well during treatment. Running heavier casing than may be required is also good practice. This gives the operator the option of fracturing down casing.

Fig. 13-5
Hydraulic fracturing a gas well.

Why do operators want higher injection rates? High injection rates produce longer fractures. As the fracture size increases, the area of formation in contact with the fracturing fluid and the corresponding fluid loss increase very rapidly. So to get the long fractures needed, the industry has gone to higher injection rates.

Sometimes, though, low injection rates are used, especially when the well is treated down tubing. This is especially true when a formation is close to a water-bearing zone. In this case, thickened fracturing fluids with good sand-carrying properties are necessary.

Fracturing materials

Fracturing fluids are categorized as aqueous base, oil base, and mixed base, depending on the main constituent of the fracturing fluid.

Aqueous-base fracturing fluids are a mixture of water and acid. Thickening agents are added to increase viscosity, which increases the sand-carrying capacity. Oil-base fluids are a mixture of oil and acid.

Emulsion-type fluids (mixed base) are made of oil and either water or acid, one phase of which is dispersed as tiny droplets into the other phase. These fluids have good sand-carrying properties with low fluid loss but are more expensive than aqueous-base fluids.

The most common propping material in the U.S. is Ottawa sand. The Canadian sand is smooth, round, and consistent in grain size. Ottawa sand is also good because it has a high compressive strength. Other types of sands are available throughout the world, but a sand of 20/40 grain size is selected for most jobs. When additional strength is required for deep-well fracturing, the engineer may use sintered bauxite.

OTHER STIMULATION METHODS

Fracturing and acidizing are the most common types of well stimulation. But several other well stimulation methods are used occasionally.

Shooting

Well shooting is accomplished by detonating charges of nitroglycerin in the well at the depth of the producing formation (Fig. 13-6). This increases the size of the wellbore and fractures the formation some distance away from the wellbore. However, the operator cannot use casing in this treatment, so open-hole completion is required.

Sometimes small shots are set off opposite the most prolific zones, especially prior to well fracturing. The theory is that shots help the fracturing materials enter the selected zones.

String shots

A string shot is accomplished by hanging a string of special primer cord down the hole opposite the zone and detonating the charges. This method may be used to remove gypsum deposits, mud, or paraffin from the face of the formation.

Reperforations

Perforations may plug after many years of producing a well. Often, it is useful to reperforate a well in the original interval if plugging is suspected. Detonating the perforating gun loosens any blocking material in the formation. This additional drainage may increase oil or gas production.

Marble shots

In open-hole completions, explosive shots sometimes are detonated with glass marbles packed around the explosives in the wellbore. When the explo-

Fig. 13-6
Nitroglycerin stimulation before (a) and after (b) stimulation.

sive charge is detonated, the marbles become projectiles, striking the face of the rock, breaking up any gypsum deposits, and perhaps even fracturing the rock. If any fractures are created by the shot, the marbles may imbed in the fracture and hold it open.

Abrasive jet cleaning

This technique uses a jetting tool with nozzles on the tubing. Streams of water or other fluid are forced through a nozzle to loosen any debris in the perforations. Some operators even inject a sand-laden fluid, which can cut through 1/4 in. steel pipe in 15-30 sec. Acid may also be applied using this tool to remove acid-soluble deposits.

Paraffin removal

Several good commercial paraffin solvents are available. The solvents can be circulated past the affected parts of the wellbore or dumped into the well to soak the accumulated foreign material. Frequently, hot oil treatments remove paraffin. Oil pumped down the tubing dissolves the deposits and carries the material back up to the surface with the production fluid.

Large-volume injection treatments

A simple treatment is just pumping large volumes of crude oil, kerosene, or distillate into the formation, especially when the formation is blocked by fine silicates or other solids. The fluids may rearrange the fine particles and open flow channels to the wellbore.

If production declines, the operator must use all available information to analyze the factors contributing to the decline. If the problem is not analyzed as completely as possible before any treatment, a great deal of money may be wasted in trial-and-error applications.

CHAPTER 14

Enhanced Oil Recovery (EOR)

Occasionally, a well will produce as much as it is able from initial, or primary, production methods. However, more oil remains in the formation. At times like these, an operator may opt to perform an enhanced recovery operation. These operations are usually divided into *secondary recovery*, or *waterflooding*, and *tertiary recovery*, commonly called EOR.

WATERFLOODING OPERATIONS

In waterflooding, water is injected into a reservoir for additional recovery. It enters the formation through injection wells in a specified pattern, depending on the individual characteristics of the formation. As the water flows from the injection wells toward the producing well, it flushes trapped oil from the formation and carries it along to the producing well. If the oil produced is equal to or greater than the amount of water produced, the well may be economical.

When determining if a reservoir is suited for waterflooding, the operator must consider the following factors:
- Reservoir geometry.
- Lithology.
- Reservoir depth.
- Porosity.
- Permeability.
- Continuity of reservoir rock properties.
- Magnitude and distribution of fluid saturations.
- Fluid properties and relative permeability relationships.

Reservoir geometry

The structure and stratigraphy of a reservoir control the location of wells and, to a large extent, dictate the methods by which a reservoir may be produced. Most water injection operations are conducted in fields with only moderate structural relief where the oil accumulates in stratigraphic traps. If dissolved-gas drive was used, much oil still remains in the formation, which makes these formations attractive for waterflooding. When all these characteristics combine together, the field is usually ideal for secondary recovery.

Lithology

Lithology includes porosity and permeability, but it also includes the rock's mineral composition. There seems to be an effect between hydrocarbons and certain types of minerals. Therefore, the formation's lithology must be studied to determine if it is suitable for waterflooding.

Reservoir depth

If a reservoir is too deep for economic drilling, or if oil wells have to be utilized as injection and producing wells, lower recoveries may be seen than where new wells can be drilled. This is particularly true in old fields where regular well spacings were not observed. Also, residual oil saturations after primary operations in most deep pools probably are lower than in shallow pools. Therefore, less oil remains.

Porosity

The total recovery of oil from a reservoir is a direct function of porosity, since porosity determines the amount of oil present for any given percent of oil saturation. Before waterflooding operations can commence, the operator needs to be sure enough space exists in the formation rock's pores to hold producible quantities of hydrocarbons.

Permeability

The magnitude of permeability of the reservoir rock controls to a large degree the rate of water injection that can be sustained in an injection well for a specified pressure at the sand face. In determining the suitability of a given reservoir for waterflooding, the operator must determine maximum permissible injection pressure from depth considerations and rate vs. spacing relationship from the pressure-permeability data. This should roughly indicate the additional drilling needed to complete the flood program in a reasonable length of time.

Reasonably uniform permeability is essential for a successful waterflood since this determines the quantities of injected water that must be handled. If great variations in permeability are noted, the waterflood will be less successful.

Continuity of reservoir rock properties

As mentioned above, uniform permeability is important for successful waterflooding. Likewise, uniformity in bedding planes (horizontal) is also important. If the beds lie relatively horizontal and there are no intrusions, the

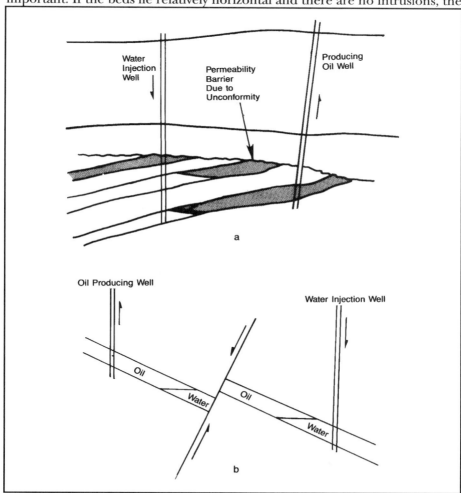

Fig. 14-1
Permeability barriers due to (a) an unconformity and (b) a fault.

waterflood operation will proceed more smoothly. However, if shale beds interrupt the smooth horizontal flow, the waterflood operation will be less successful (Fig. 14-1).

Magnitude and distribution of fluid saturations

Usually, high oil saturation is more suitable than low oil saturation in waterflood operations. The higher the saturation, the higher the recovery efficiency. Also, ultimate recovery will be higher, water bypassing will be lower, and the economic return per dollar risked will be greater. When there is less connate water left in the formation, the operator can pretty well guess that what remains behind will be oil.

Where can the operator find out this information? From cores taken from newly drilled wells. He can also confirm his guess using information from electric log, laboratory oil floods, and capillary pressure tests.

Fluid properties and relative permeability relationships

The physical properties of the reservoir fluids also have pronounced effects on the advisability of waterflooding a given reservoir. Viscosity is one of the most important properties because it affects the mobility ratio. Relative permeability is also important. The larger the mobility ratio, the lower the recovery breakthrough and hence the more expensive the waterflood operation will be.

Water sources

One of the major considerations in waterflooding is where to locate enough water for the operation. During the early injection life of the reservoir—the fill-up period—a high rate of 1-2 barrel per acre foot is needed. After initial fill-up, the rate decreases to less than1 barrel per acre foot. Ultimately, the volume of water will be about 150-170% of the total pore space in the formation, which needs to include the pore volume of any adjacent sands.

Operators can choose to use salt water or fresh water, depending on the operation. Where economics permit, salt water usually is preferred to fresh water. In most fields, salt water formations exist above or below the oil zones. Wells can be dug to the appropriate depth, and water can be pumped to the surface and reinjected into the appropriate injection wells. If the operation is near the ocean, sea water can be used. However, it must sometimes be treated.

Fresh water can be found from surface sources like ponds, lakes, streams, and rivers when these do not interfere with local water requirements. However, these sources may have limited capacity during drought periods and the water often needs expensive treating. A more favorable method is to use alluvial beds near rivers. Shallow wells are dug into these formations; the only major draw-back is that the water must be treated for bacteria. Finally, fresh water can be located below the surface, usually anywhere from ground level down to 1,000 ft. In this case, wells must be dug and pumps installed. Again, the economics must be weighed against the profits for the waterflood project.

Waterflood patterns

We mentioned earlier that a dominant factor controlling waterflood oper-ations is the location of injection and producing wells. Most of this is based on complex flow geometry. For purposes of this book, however, we should know that there are a few basic patterns that operators use frequently during water-flood operations.

When reservoirs are continuous and have relatively large areal extent, the patterns form a symmetrical and interconnective network of wells (Fig. 14-2). Four basic patterns are usually chosen: direct line drive, staggered line drive, five-spot pattern, and seven-spot pattern. Sometimes, though, it is impossible to work around one of these standard patterns, so the operator must modify them.

Water injections wells are another important consideration in choosing a pattern. The rate of injection depends on effective permeability, oil and water viscosity, sand thickness, effective well radius, reservoir pressure, and applied water pressure. These factors influence how many wells must be drilled and in what geometrical pattern.

Occasionally, *pilot floods* are conducted. These floods evaluate operational procedures and give advance information on how extensive the waterflood will perform and whether the proposed configuration, or pattern, will be the best choice. In an extensive pattern flood, the perimeters of the well flooding pat-terns act as impermeable boundaries. However, in a pilot flood only one or two well patterns within the reservoir are displayed. Therefore, production is defi-nitely different, but the operator has a chance to test waterflooding potential.

Water treating

As we've discovered, water is one of the most important components in a waterflood operation. During the early days, operators only considered quan-tity, not quality. Today, however, operators know that poor water treatment can do as much damage to a waterflood project as any other factor.

Once a water source is chosen, the operator conducts a water analysis to determine the following:
- Compatibility with the reservoir water.
- What kind of injection facility is best suited.

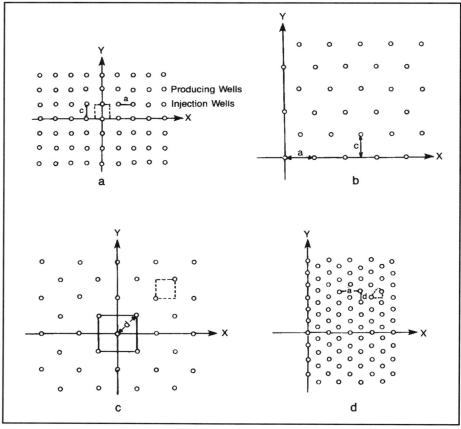

Fig. 14-2
The four basic types of waterflood patterns: (a) direct-line-drive pattern, (b) staggered line drive pattern, (c) five-spot pattern, and (d) seven-spot pattern.

- The treatment necessary to have an acceptable water for the reservoir with minimal corrosion of equipment.

This analysis should be conducted periodically to also determine the presence of three undesirable constituents of water: dissolved gases, minerals, and microbes.

Precipitation is one of the main problems with minerals. When the minerals precipitate out of solution, they close up the pores in the rock and diminish porosity in the formation. *Sequestering and chelating agents* are added to the water to help prevent this precipitation.

"Sequestering" means to set apart or to separate. "Chelating" pertains to a group which attaches itself to a central metallic atom by means of two valences to form a heterocyclic ring. In other words, the sequestering agent separates the metallic cation from the anion by chelation, which cures the precipitation problem.

Corrosion inhibitors are also often added to water. These are chemicals that control the corrosive activity between the metallic alloy and water. The advantage in using these is that the piping and tubing do not wear out as quickly and maintain strong production rates.

Residual oil after waterflooding

Once a reservoir has been waterflooded, is all the oil washed or flushed from the formation? Unfortunately, no. Much of it remains behind and cannot be produced by water alone. Engineers can determine the amount by using cores or by studying the results of waterflood susceptibility tests on a representative sampling of the reservoir rock. With either test, some amount of residual oil is estimated. If an operator determines that there is still an appreciable amount of producible hydrocarbons downhole, he may opt for a third stage of production: tertiary, or enhanced, oil recovery.

ENHANCED OR TERTIARY RECOVERY METHODS

After all the possible oil has been produced from the well using primary and secondary operations, an operator may decide to try yet a third type of recovery technique: enhanced or tertiary recovery. These methods are very expensive, and many are still in experimental stages. However, they do increase the production from oil wells and may someday be more economical if the price for a barrel of crude rises high enough.

There are three general classifications of enhanced oil recovery, or EOR, methods that show significant promise: chemical flooding, miscible flooding, and thermal recovery. Within these three general methods, six distinct processes are considered economically feasible: polymer flooding, surfactant flooding, alkaline flooding, CO_2 flooding, steam injection, and in situ combustion. Let's look at these EOR methods and learn more about how they work.

Chemical flooding

Chemical methods include polymer flooding, surfactant (micellar-polymer, microemulsion) flooding, and alkaline flooding processes. These processes are broadly characterized by the addition of chemicals to water in order to generate fluid properties or interfacial conditions that are more favorable for oil production. Three techniques are widely used: polymer flooding, surfactant flooding, and alkaline flooding.

Fig. 14-3
Polymer flooding (adapted from original drawings by Joe R. Lindley, U.S. DOE, courtesy National Petroleum Council).

Conventional waterflooding can often be improved by adding *polymers* to the injection water to improve, or decrease, the mobility ratio between the injected and in-place fluids (Fig. 14-3). In other words, the polymer makes it easier for the hydrocarbons to move through the formation. The method is usually used on reservoirs whose production capacity extends over a large area. This is because the polymer solution sweeps a large fraction of the reservoir, not just the water alone.

Surfactants, or surface active agents, can also be added to water. These chemicals reduce the forces that trap the oil in the pores of the rock. A surfactant slug displaces the majority of the oil from the reservoir, forming a flowing oil-water bank that moves ahead of the surfactant slug. Following the surfactant slug is a slug of water containing polymer in solution. The polymer improves the sweep efficiency and helps ensure that as much oil as possible has been removed from the pores. The polymer solution slug is then followed with a plain water slug. The process may then be repeated until the reservoir is sufficiently "clean."

Alkaline flooding uses inorganic alkaline chemicals such as sodium hydroxide, sodium carbonate, or sodium orthosilicates in water to enhance recovery

Fig. 14-4
Alkaline flooding (adapted from original drawings by Joe R. Lindley, U.S. DOE, courtesy National Petroleum Council).

by reducing interfacial tension, altering the wettability, or causing spontaneous emulsification (Fig. 14-4). The process is less costly then surfactant flooding, but it has a lower recovery potential, too.

Miscible methods

Miscible floods usually use carbon dioxide, nitrogen, or hydrocarbons as solvents to increase oil production. Some of these kinds of floods have been in operation since the 1950s.

Although CO_2 flooding is relatively recent, it is expected to make the most significant contribution to miscible enhanced recovery in the future (Fig. 14-5). Carbon dioxide is a powerful mover with oil. Initially, the CO_2 is not miscible, or mixable, with the oil. However, as it contacts the crude oil in the reservoir, it extracts some of the hydrocarbon constituents of the crude oil and becomes dissolved into the oil. When the oil and CO_2 mix, you have the same sort of phenomenon as you do with gas lift: the oil thins and can move more easily.

Fig. 14-5
CO_2 miscible flooding (adapted from original drawings by Joe R. Lindley, U.S. DOE, courtesy National Petroleum Council).

For some reservoirs, miscibility between carbon dioxide and oil cannot be achieved but CO_2 can still be used to recover additional oil. The gas still swells within the reservoir and reduces the oil's viscosity. Together, these qualities improve the mobility of the oil.

Hydrocarbon gases and condensates are also used in miscible flood projects. Generally, these light hydrocarbons are too valuable to be used commercially, so the processes are expensive. Nitrogen and flue gases are also used, but they are often effective only in high-pressure, high-temperature wells.

Thermal methods

Thermal EOR processes add heat to the reservoir to reduce oil viscosity and/or to vaporize the oil. In both instances, the oil is made more mobile so that it can be driven more effectively to producing wells. In addition to adding heat, these processes provide a driving force (pressure). There are two principal thermal recovery methods: steam injection and in situ combustion.

Steam injection generally occurs in two steps: steam is pumped down a producing well to heat and thereby loosen oil near the producing wellbore, and

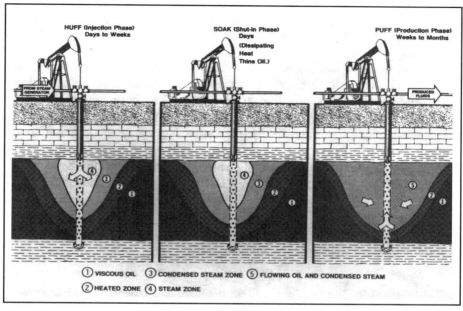

Fig. 14-6
Cyclic steam or huff & puff stimulation method (adapted from original drawings by Joe R. Lindley, U.S. DOE, courtesy National Petroleum Council).

steam is driven down an injection well and moves through the reservoir toward a producing well, pushing the warm, movable oil in front of it.

In actual practice, a mixture of steam and hot water is injected into the formation. Normally, steam is generated at the surface, but some heat is lost and part of the steam may turn to hot water before it reaches the producing formation. When this steam/hot water mixture is used in cycles at a producing well, the technique is called huff and puff or steam soak (Fig. 14-6).

In situ combustion is normally applied to reservoirs that have low-gravity oil, but it has been tested over a wide spectrum of conditions. Heat is generated within the reservoir by injecting air and burning part of the crude oil. This reduces the oil's viscosity and partially vaporizes the oil in place. Then the oil is driven forward toward the producing well by a combination of steam, hot water, and gas drive.

We need to remember that tertiary recovery techniques are very expensive, especially when the price for a barrel of crude oil is low in the marketplace. Waterflooding is usually more economical, so it is used more frequently. Also, these methods are used to recover crude oil, not natural gas. Let's finish our discussion of production by studying natural gas further and learning how it is processed.

CHAPTER 15

Natural Gas Processing and Cogeneration

Natural gas processing and cycling plants recover salable liquids from gaseous streams produced directly from gas wells or from normal oil and gas separation equipment on oil wells. Such plants vary greatly in size and capacity, from several million to several hundred million cubic feet of gas per day.

The use of conventional oil and gas separators and emulsion treaters is not considered gas processing, nor are treatments to remove contaminants such as dust, dirt, water vapor, hydrogen sulfide, and carbon dioxide from the gas. These treatments are normally referred to as *gas conditioning*. Gas processing, on the other hand, is any operation whose primary purpose is the recovery of liquids from gas.

Cycling plants are used primarily on gas/condensate reservoirs. With these reservoirs, maintaining reservoir pressure above the dew point—the pressure point below which liquids will form—is desirable to increase recovery. After the heavier hydrocarbons have been removed in liquid form, all or part of the remaining "dry" gas is pumped back into the reservoir to help maintain reservoir energy. This, plus the economic reasons, sometimes make gas processing a profitable operation.

NATURAL GAS TERMINOLOGY

The terminology used to determine natural gas streams is colorful but hardly precise. Consider, for example, the classifications of gas as "wet," "dry," "rich," and "lean." *Rich* or *wet* gas usually means a stream that is potentially worth processing for its liquid. *Dry* or *lean* gas denotes the opposite. In short, these terms are about as quantitative as the terms "fat" and "skinny" applied to people.

The quantitative yardsticks for natural gas streams are gal/Mcf (gallons of condensable liquid per 1,000 cu ft) and the percentage content of various

Table 15-1
Typical Natural Gas Components

Hydrocarbon	Amount, %
Methane	70 - 98
Ethane	1 - 10
Propane	trace - 5
Butane	trace - 2
Pentane	trace - 1
Hexane	trace - 1/2
Heptane +	none-trace
Nonhydrocarbon	
Nitrogen	trace - 15
Carbon dioxide	trace - 1
Hydrogen sulfide	trace occasionally
Helium	trace 5

chemical components. Gal/Mcf can be derived by calculations based on chemical analysis or from standard compression or carbon absorption tests.

The term *casinghead gas* usually refers to gas obtained from a well in association with crude oil. *Residue gas* is any gas suitable for sale as commercial natural gas that comes from a processing plant. The term implies that all readily liquefiable components have been reduced to satisfactory proportions.

Sweet gas means that the content of hydrogen sulfide, other sulfur compounds, and CO_2 is low enough that the gas may be sold commercially without further effort to remove these compounds. *Sour gas* denotes the opposite.

TYPES OF NATURAL GAS LIQUIDS

Natural gas liquids are hydrocarbons that have the properties of the normal paraffins ethane through decane with no more than trace amounts of anything heavier. Natural gas contains hundreds of hydrocarbon compounds, most of which are present only in trace quantities. The liquids normally are described in terms of their primary chemical content, boiling point, vapor pressure, color, purity, and certain other qualities. Various natural gas products, as described below, are defined according to arbitrary standards established by the Natural Gas Processors Association (NGPA).

Commercial propane

This is a hydrocarbon product composed predominately of propane and/or propylene and having a vapor pressure of not more than 215 psig (pounds per square inch gauge) at 100° F. It must consist of at least 95%

propane and/or propylene. Also, it must pass NGPA tests covering total sulfur content, corrosive compounds, dryness, and the amount of residue left on evaporation in a standard test.

Commercial butane

Like commercial propane, commercial butane must meet the same general tests for contaminating substances. It is a hydrocarbon product containing predominately butanes and/or butylenes and having a vapor pressure of not more than 70 psig at 100° F. At least 95% must evaporate at a temperature of 34°F or lower in a standard test.

Liquefied petroleum gas (LPG)

LPG is a mixture of commercial propane and commercial butane. The maximum vapor pressure cannot exceed that of commercial propane, and the residue on evaporation cannot be greater than commercial butane. The specific mixture is designated by its vapor pressure in psig at 200° F. The actual vapor pressure of a mixture so specified must not vary by more than 0-5 psi from its designated pressure. So a 100-psi LPG must have a vapor pressure of at least 95 psi but not more than 100 psi.

Natural gasoline

This petroleum product is extracted from natural gas and falls within the following specifications:
* Vapor pressure: 10-34 psi.
* Percentage evaporated at 140° F: 24-85%.
* Percentage evaporated at 275° F: not less than 90%.
* End point (distillation): Not higher than 375° F.
It must also pass specified corrosion, color, and sourness tests.

Vapor pressure, usually referred to as *Reid vapor pressure* (RVP), is a standard test result used to designate grades. For example, motor fuels are usually 5-8 psi RVP mixtures. Very light oils (60-70° API) have a 12-psi RVP. Most marketers specify a natural gasoline product of 14-26 psi RVP. But because of the increasing use of this fluid as refinery blending stock, there is a trend toward specifying not more than 18-psi RVP.

Ethane

Ethane is in growing demand as a base for the manufacture of plastics, alcohols, and other chemicals. Normally it is produced as a separate fluid only in large plants because of the high capital investment required. Most ethane-producing plants are located near petrochemical installations. In spite of the rapid growth of the market for this fluid, most processing plants still sell ethane as part of the natural gas.

METHODS OF PROCESSING NATURAL GAS

There are several ways in which the liquefiable components may be removed from natural gas: absorption, adsorption, refrigeration, and combinations of these.

Absorption

This is a process in which rich gas is contacted with a heavier hydrocarbon oil that has been fractionated to specification (Fig. 15-1). The necessary contact is provided by flowing the crude oil down through a contact tower known as an absorber, countercurrent to the gas flowing upward. The amount of liquefiable gas components absorbed by the oil depends on the pressure and temperature of the absorber, the relative flow rates of the gas and oil, and the inlet composition of the gas and oil, as well as the amount of contact achieved.

Regular absorption plants perform this process at ambient temperatures, which generally are 80-120° F. Most plants are designed to recover 40-75% of the propane in rich gas. Recovery of butanes and heavier hydrocarbons depends on the amounts present initially and their molecular weight. Recovery of hexanes and heavier hydrocarbons, for example, approaches 100%.

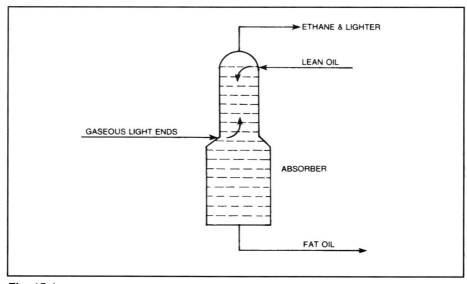

Fig. 15-1
Schematic of an absorber.

Refrigeration-absorption

Refrigeration-absorption plants operate at low temperatures by refrigerating both inlet oil and gas to temperatures from -10 to 20° F. Most plants designed for high-efficiency propane recovery (up to 85%) now use this approach. With the exception of the temperature difference and the facilities for supplying it, this plant is essentially the same as a regular absorption plant.

Refrigeration

These plants are based on the principle that the lower the temperature, the greater the tendency for a component to liquefy. Around 0° F, 40-75% of the propane will condense out, along with proportionately higher percentages of the heavier components. This type of plant is simpler and thus finds widest application on very rich gas streams and where the quantity of gas is limited.

Adsorption

The widest application of adsorption plants is relatively small, fairly lean gas streams where processing will be performed at the lease. Most of these plants also dehydrate the gas to meet pipeline specifications for water vapor content.

ECONOMICS AND CONTAMINANTS

Virtually all contracts require that gas be almost free of any liquids that might condense in the transmission line, to be "sweet," and to have a water content limited to 7 lb/Mcf. In almost all cases, gas conditioning that satisfies the contract also meets the limitations imposed by the liquid recovery system. When determining if a plant will be profitable, this factor must be considered.

An increasing number of gas and gas condensate reservoirs are being discovered that contain hydrogen sulfide (H_2S) or CO_2 in such quantities that extensive treatment is required. Consequently, larger and more complex facilities are often needed to meet the specifications of the gas sales contract. Many plants contain provisions for converting the recovery hydrogen sulfide to sulfur and recovering the CO_2 for commercial use. In such cases the cost of conditioning the sour gas before processing can be a substantial part of the total plant cost.

Sulfur is a very low-priced commodity, so the feasibility of converting H_2S to sulfur depends to a large measure on the availability of other sources of sulfur in the area. The cost of transporting raw sulfur for use in chemical operations is likely to be prohibitive unless the consumer is close to the gas processing plant or inexpensive transportation is available.

Natural gas processing is an important part of the production process. Years ago, the gas was just vented to the atmosphere or flared—virtually wasted. Today, we learn to make use of this valuable natural resource. Much the same principle is being used in one of many exciting new alternative energy methods: cogeneration.

COGENERATION

Cogeneration—producing two forms of energy such as process heat and electric power from a single fuel—offers the petroleum industry a major new growth area, whether to improve gas demand or process plant economics, or as an investment in a larger role as an electrical power producer. The scramble for that market, pegged at about $16 billion this century, is under way.

A cogeneration plant is a coal or gas-fired (occasionally biomass such as peat or wood chips are fuels) plant that generate both process (commercial) steam and electricity for in-plant use or for sale. It's easy to see how some of that expensive-to-produce natural gas could be converted to another energy source and either used at a processing plant or sold back to local utility companies.

Beyond boosting gas demand and improving process plant profits, cogeneration can be a boon in other areas of the oil patch.

Major firms with huge heavy-oil reserves in California's San Joaquin Valley, for example, are looking to cogeneration to spur big steamflood projects or expansions (Fig. 15-2). Such projects often have been stymied in the past because air emissions from burning lease crude to generate steam violated regional standards. But burning lease crude rather than cleaner-burning gas

Fig. 15-2
Cogeneration is being used to help steamflood operations in heavy-oil reservoirs.

Fig. 15-3
Typical cogeneration power plant in a Kern County, Calif. heavy oil field (photo by
Iger, Studio courtesy ARCO).

was cheaper because the operator would not have to pay a royalty nor a tax on the produced oil since it was burned and not sold.

Cogeneration power sales help offset gas fuel costs, enabling operators to replace crude with gas to generate steam. Then they can bank the emissions offset credits earned with the substitution and save them for expanded projects (Operators are allowed limited emissions that can be transferred to other projects). That approach should lead to a significant increase in heavy-oil production this century.

Cogeneration will benefit utilities over the long run because it will recapture the gas-fired power generating market they have lost while avoiding the cost of replacing or building new base load plants. But there are limits to any restructuring of the utility industry. If too much cogeneration comes on stream, the avoided cost will come down, and many cogeneration projects will no longer be economically feasible. Moreover, much of cogeneration's current appeal stems not only from the public's energy-saving attitudes instilled by oil price shocks but also from objections to nuclear and other traditional central power plants.

For the near term, the only significant growth in power generating capacity will be in cogeneration and other forms of independent power production. Still, cogeneration will be a sustained market because there will be more growth in the overall electrical power market than cogeneration alone can meet.

CHAPTER 16

New Technology

Oil and gas companies continually press the envelope of technology in their unending campaign to optimize production rates and ultimate recovery of oil and gas reserves while at the same time keeping a lid on costs and unwanted environmental effects to a minimum.

Some of these new technologies are really not new at all but are simply new applications or enhancements of existing technology. Some of these enhancements involve dramatic advances in computer technology, an area that is certain to yield still more advances in the technology of oil and gas production.

This chapter offers an overview of some of the more significant recent advances in petroleum production technology and is by no means intended to be comprehensive. There are entire volumes devoted to specific new technologies and cutting-edge applications of existing ones, but the following sections cover some technologies that are getting most of the attention in production circles today.

HORIZONTAL/EXTENDED-REACH DRILLING

The petroleum industry has been producing oil and gas through inclined and/or horizontal wellbores since the 1940s. However, prior to 1979, very few wells had been drilled. Until then, the industry generally relied on techniques such as hydraulic fracturing to boost productivity of vertical wells.

While horizontal or extended-reach wells offer much greater production rates than do vertical wells because they expose more of the producing formation to the wellbore (Fig. 16-1), the greater drilling and completion expenses associated with them generally restricted their use to limited applications. That changed with the dramatic advances in drilling technology, especially the innovations in measurement while drilling technology, which is covered in the next section.

With a better understanding of the technology involved, more operators now have the incentive to choose candidates for horizontal/extended-reach

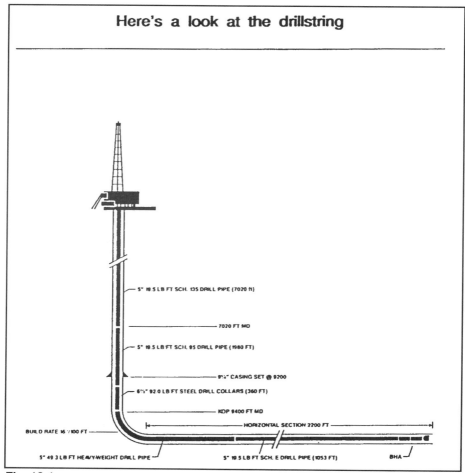

Fig. 16-1
Typical horizontal well.

drilling and drill, complete, test, stimulate, and produce from reservoirs via inclined or horizontal wellbores that sometimes yield exponential increases in well revenues. Accordingly, there is a strong push under way around the world to drill horizontal wells.

Extended-reach wells can be classified as having ultrashort, short, medium, and long turning radiuses before becoming horizontal (Fig. 16-2).

An ultrashort turning radius horizontal well, or drainhole, involves a turning radius of 1-2 ft. The drilling process calls for underreaming the vertical

TYPES OF HORIZONTAL WELLS

A. DRAINHOLE OR
ULTRASHORT TURNING
RADIUS OF
1-2 FT
(0.3-0.6m)

B: SHORT TURNING
RADIUS OF
20-30 FT
(6-12m)

C: MEDIUM TURNING
RADIUS OF
300-400 FT
(91-152m)

D: LONG TURNING
RADIUS OF
600-2000 FT
(183-610m)

Fig. 16-2
Types of horizontal wells (courtesy Halliburton Energy Services).

hole to at least a 24 in. diameter for about 6-10 ft in the vertical section. A water jet then drills a 1.5-2.5 in. diameter hole to a length of 100-200 ft.

A short turning radius hole has a radius of 20-40 ft, and the wellbore length is about 200-700 ft. The first cut following the initial vertical drilling is a 15-20 ft window in the vertical casing, and the wellbore is kicked off through the window. A whipstock and curved drilling entry guide assist the flexible drill collars in making the short turning radius before going horizontal.

A medium turning radius horizontal well has a radius of 300-500 ft, and the horizontal section usually is as much as 1,500 ft long. These wells are typically drilled with a downhole mud motor and flexible drill pipe. "Build" motors are used to build the wellbore angle at a rate of 20°/100 ft. The horizontal sections are then drilled with "angle hold" motors.

A long turning radius horizontal well has a radius of 600-2,000 ft and relies on conventional drilling tools. A combination of drill bits with bent subs and down-hole mud motors are used to drill horizontal sections that can exceed 4,000 ft.

Fig. 16-3 shows that a horizontal well with a length of 1,000 ft placed in the center of a formation will produce tenfold the volume of a vertical oil well in 100 days. That same well will produce twice as much as a vertical well over a period of 5 years. So it follows that more oil will ultimately be recovered and recovered much sooner by a horizontal well (provided the reservoir parameters are suitable for a horizontal well). The much quicker and bigger payout thus justifies the added expense of drilling and completing a horizontal well.

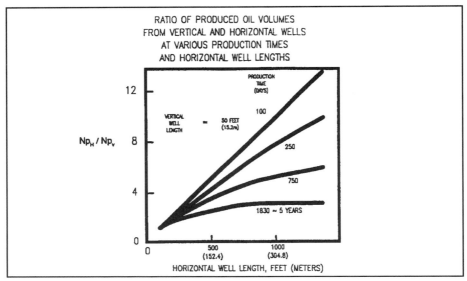

Fig. 16-3
Production from horizontal vs. vertical wells (courtesy Halliburton).

The kinds of formations that benefit from a horizontal or extended-reach well are generally the same that would benefit from a controlled induced hydraulic fracture: thin formations, some naturally fractured formations, low permeability formations, formations with a gas cap and/or bottom water, some layered formations, and partly depleted formations.

Horizontal and extended-reach wells carry some unique problems. Exposure of more of a formation to drilling fluids for a longer time could result in formation damage that is more pronounced in horizontal wells than in vertical wells.

But because of the greatly improved return on investment that horizontal and extended-reach wells offer compared with vertical wells in selected instances, there is little doubt that they are here to stay.

MEASUREMENT WHILE DRILLING

Measurement while drilling (MWD) entails using specially designed electronic tools that impart real-time well status data from downhole to the surface, hence the name. The idea is to maximize drilling precision and efficiency and reduce the instances of downtime and costly service jobs that result from such problems as lost circulation material, correcting the drilling azimuth, and drill bit damage.

MWD tools typically transmit infromation to the surface by encoding data as series of pressure pulses in the mud column. An MWD system such as Halliburton's consists of three fully integrated assemblies: a pulser, a battery power pack, and a sensor system.

On the surface, MWD equipment can include: surface sensors to detect the mud pulses, with the option to monitor depth and drilling parameters; a computer system that decodes transmitted data and presents it as real-time data; an application computer system that analyzes data such as computer-aided directional drilling and survey software and magnetic azimuth correction algorithms

COILED TUBING

Coiled tubing is another technology that has been around for decades and seen limited applications until significant advances resulted in a recent surge of interest and use.

The coiled tubing unit is a self-contained, easily transported, hydraulically powered workover unit that injects and retrieves a continuous string of tubing into a larger string of tubing or casing (Fig. l6-4). The system can be used on land or offshore and does not require a separate workover rig. The unit can be used on live wells and allows continuous injection of fluids or nitrogen while continuing to move the pipe.

The primary advantage of coiled tubing technology is that is a cost-effective substitute in many instances for costly workover jobs by service rigs. Coiled tubing operations can involve applications for drilling, remedial cementing, electric wireline logging and perforating, fluid displacement, sand control, wellbore cleanout, setting and retrieving bridge plugs, permanent installations, running packers, using coiled tubing as flowline, securing flow control of horizontal and deviated completions, fishing tools, testing, completions, well control, and stimulation.

Coiled tubing drilling is becoming an increasingly viable option for many operations. Among its applications are exploratory wells, wellbore extensions, and horizontal reentry from an existing vertical well.

With coiled tubing drilling, a continuous piece of tubing is used to deploy a drilling device, usually a positive displacement motor, to the drill target. Hydraulic power to turn the drill bit is provided by surface pumps.

By installing an electric wireline inside coiled tubing and using the tubing as a conduit, many standard wireline services can be performed in highly deviated and horizontal wellbores. Coiled tubing-conveyed wireline services include production logs, cement bond logs, casing inspection logs, and perforating.

Fig. 16-4
Coiled tubing operation under way.

Coiled tubing applications involving fluid displacement usually are in conjunction with nitrogen, which is beneficial in developing and applying advanced fluid systems to achieve many reservoir management objectives.

Conveying perforating guns to the target zone with coiled tubing has been used effectively for many years in a number of applications, offering faster run-in times, allowing perforating in underbalanced well conditions and in horizontal wells where conventional perforations are impractical, and using the coiled tubing as a production string.

Coiled tubing offers advantages when placing resin material for sand control by reducing interface dilution, contamination, and spacer requirements. By placing the end of the coiled tubing directly at the perforations, this permits sand entry into the perforations almost immediately.

In well stimulation applications, coiled tubing is the most effective method of placing treatment fluids precisely at the target zone. Using coiled tubing for stimulation applications also protects production tubing from treatment fluids and avoids contamination of acid by production tubing deposits and scale, enhancing acidizing success. A coiled tubing unit also can inject paraffin and corrosion inhibitors.

Coiled tubing benefits include: improved efficiency because a well does not have to be killed to be serviced; greatly reduced rig- up and trip times; and reduced potential for formation damage because it avoids using kill fluids in any operation.

Halliburton Energy Services also cites these environmental benefits of coiled tubing applications when compared with a conventional drilling or service rig: a reduced wellsite area with fewer pieces of equipment; a smaller volume of drilling fluid because the coiled tubing diameter is smaller than conventional tubing; reduced noise level; reduced visual profile with elimination of a derrick; reduced damage to roads because coiled tubing units use about one-tenth the number of truckloads that is required to transport a conventional rig; and smaller volume of formation cuttings that require disposal.

THREE-DIMENSIONAL SEISMOLOGY

Of all the technical advances in seismology during the past two decades, none has approached in effectiveness the extension from two-dimensional to three-dimensional seismics. The reasons are as follows: Seismic waves propagate in three dimensions, and subsurface structure invariably has some three-dimensional complexity. The power of two-dimensional seismics for imaging the subsurface is, therefore, restricted—often severely so.

Increased density of spatial coverage

Where 2-D data are obtained along isolated lines, 3-D data are acquired and processed such that the resulting stacked and migrated traces are closely spaced on a rectangular grid spanning the prospect area.

The high density of regularly spaced traces provides the interpreter with detailed information about the full 3-D subsurface volume. Any desired cross-section can be extracted from the volume for display and analysis. Therefore, the interpreter can analyze closely spaced cross-sections in either principal directions of the rectangular grid without having to rely upon smooth interpolation to span large distances between data control.

From the 3-D data volume, one can also readily extract vertical sections along any desired zigzag path, for example, a loop connecting borehole locations. In fact, sections extracted for analysis need not be vertical.

Comprehensive overviews of the structure, in particular, faulting, can be gained by slicing the data volume horizontally at fixed intervals. Likewise, seismic attributes such as amplitude, dominant frequency, and derived interval space can be mapped and displayed along curved reflector surfaces.

Enhanced subsurface imaging

In addition to the enhanced lateral detail available from the dense spatial coverage in 3-D surveys, dramatic improvement in the quality of subsurface imaging is made possible by 3-D migration. With 2-D survey data, migration is at best, incomplete.

The process cannot comprehend energy reflected by scatters outside the plane of the survey line. The great advance from 2-D to 3-D seismic is achieved by the 3-D migration process, made possible only when the dense coverage of 3-D data is obtained over a survey area. The improvement in subsurface imaging obtained by 3-D migration of 3-D survey data is no less important where the target is strategic, in mostly flat-lying bedding, than where it is highly structural. Furthermore, proper preservation of amplitude, essential for inversion to obtain thin layer models of porosity requires the accurate positioning of events that is most closely achieved through 3-D migration.

FIELD DELINEATION, MANAGEMENT, AND MONITORING

Historically, the seismic method has been primarily a tool for hydrocarbon exploration. Its role has been to reduce the risk of finding new fields. Most often, today, the greatest cost in developing petroleum reserves is incurred after the discovery well has been drilled. With its ability to image the subsurface with increased accuracy and sharpened detail, 3-D seismic is now seen as a key tool for delineating the boundaries of fields and for analyzing field heterogeneity.

Structural interpretation and lithologic modeling of migrated 3-D data enable seismically based interpolation of well log data, yielding accuracy and detail in reservoir description, for reservoir simulation and, in turn, reservoir management.

3-D surveys are now being used to monitor the progress of different types of enhanced recovery projects, such as water flooding and steam flooding, as well as to identify locations for proposed injection wells. Their use is supported by the economics of reservoir description, field development, and field management. The cost of a 3-D survey can be a small fraction of the cost of the production and injection wells required to develop a field.

While 3-D seismic is not a new idea, dramatic advances have been made recently in the efficiency and comprehensiveness of data processing, display, and interactive interpretation. These advances have made the 3-D seismic method a cost effective tool for optimizing field development and management.

CHAPTER 17

The Future

Having learned about oil and gas production, it would be worthwhile to speculate about the future of the oil and gas producing industry.

For almost as long as this industry has existed, there have been repeated warnings about its imminent demise. There have been estimates, from respected authorities, of a rapidly disappearing reserves base, with the sense that the world is quickly running out of oil (Gas has never seemed to have this problem). Each time, the predictions proved dramatically wrong. It is perhaps the unintended consequence of these warnings that the speculative value of a resource perceived to be dwindling was enough to spur the innovators and risk-takers that have always dominated this industry to go out and find still more oil.

One thing that the oil price shocks of the 1970s proved is that the perceived scarcity of petroleum inevitably leads to higher prices, which in turn leads to the incentive to find more petroleum. Even while the U.S. and other nations embarked on crash courses of developing alternate and synthetic fuels and promoting energy efficiencies, the incentive was there to find and produce still more oil.

Just before the oil supply crises, at the start of the 1970s, total world proved reserves of crude oil averaged about 600 billion bbl. By the time of the oil price collapse in 1986, that number had jumped to more than 700 billion bbl. That was in an environment when oil prices were significantly higher in real and nominal terms than they are today. Since 1986, with oil prices often at historically low levels in real terms, total oil reserves have jumped to more than 1 trillion bbl. That's more than a 40% increase in reserves added in less than a decade when prices generally were depressed!

And what has happened to the countermeasures taken in the belief that oil was running out? Conservation measures have reached the limit of their cost-effectiveness, and improvements in energy use per capita in the industrialized nations have pretty much stalled out. Alternate energy sources such as solar and wind energy, synthetic fuels such as shale oil and biofuels, and new energy technology such as fusion and hydrogen fuel remain promises for the distant future or costly boondoggles. The much-touted synthetic fuels industry the U.S. government tried to create by fiat during the early 1980s became a white elephant that remains a symbol of wrongheaded government intrusion into energy markets.

One thing the price collapse taught the industry is that a perception of increasing scarcity and ever-rising prices is a myth. Sooner or later, supply and demand come back into balance, and the price of oil and gas is restored to something approaching its actual replacement level.

The latest prophecy of doom for the petroleum industry comes from the environmental lobby, which continues to predict environmental Armageddon for the earth if such a heavy reliance on non-renewable fuels continues. It is no secret that the petroleum industry is the favorite target of environmentalists, mainly because it is an easy target as a result of its involvement with what is arguably the world's most critical producible commodities and a negative image spawned by such atypical occurrences as oil price shocks and catastrophic oil spills.

The decade of the 1990s became known as the Decade of the Environment, ushered in by the 1989 Exxon Valdez oil spill off Alaska. Because of the growing burden of environmental regulations affecting the oil and gas industry that multiplied still more after the Exxon Valdez spill, many in and out of the industry saw a revived environmentalism as signaling a death knell for petroleum.

The dangers to the industry from extreme environmental measures remains. A study by the American Petroleum Institute in early 1995 found that new federal environmental requirements under consideration could cost about 19,300 jobs in the U.S. exploration and production industry, along with about 35,200 indirect jobs. In addition, the cost for industry to comply with the regulations could reach $9.6 billion initially and climb to $14.2 billion in 5 years. API claims the proposals on air, water, chemical releases, underground injection control, and production waste disposal could prompt the shut-in of 171,000 oil wells and 33,000 gas wells in the U.S.

However, the trend in the U.S. and in many other countries today is toward less regulation, not more; less intrusion by government in the marketplace, not more. The current experiment with introducing reformulated gasoline in the U.S. to ostensibly benefit air quality in certain highly polluted areas illustrates that many people will resist new government interference in petroleum markets even if it is touted as environmentally beneficial, once they start feeling the results in their pocketbooks.

This is not a cynical assertion; rather, it presupposes that when an uninformed public, already alarmed about the growth and scope of government, discovers unacceptably high costs associated with government measures, it

Table 17-1
EIA's World Oil Consumption and Production Base Case

OTHER FORECASTS

ENERGY INFORMATION ADMINISTRATION
INTERNATIONAL ENERGY OUTLOOK—1993
WORLD OIL CONSUMPTION AND PRODUCTION BASE CASE
(Million Barrels per Day)

Supply and Disposition	History			Projections		
	1989	1990	1991	1995	2000	2010
Market Economies						
Production						
United States	9.88	9.68	9.88	9.0	8.3	8.6
Canada	2.03	2.02	2.04	2.1	2.1	2.2
OECD Europe	4.38	4.54	4.78	6.2	6.2	4.6
OPEC	23.81	25.10	25.41	28.4	33.1	42.7
Other countries	10.43	10.80	11.05	13.1	13.6	12.2
Net CPE Exports	2.09	1.88	1.27	0.9	0.7	0.0
Total	52.62	54.02	54.43	59.5	64.0	70.3
Consumption						
United States	17.33	16.99	16.71	18.2	19.3	21.0
U.S. Territories	0.21	0.21	0.24	0.2	0.3	0.3
Canada	1.73	1.69	1.61	1.7	1.8	1.8
Japan	4.98	5.14	5.29	5.7	6.1	6.5
Australia & New Zealand	0.79	0.82	0.80	0.8	0.9	1.0
OECD Europe	12.83	12.91	13.33	13.9	14.5	15.0
Other market economies	14.92	15.69	16.30	19.4	21.5	25.0
Total	52.79	53.45	54.28	59.8	64.3	70.6
Discrepancy	0.22	−0.07	−0.22	0.3	0.3	0.3
Centrally Planned Economies (CPE)						
Production						
China	2.76	2.77	2.80	2.9	3.1	3.5
Soviet Union	12.14	11.40	10.41	8.4	9.3	11.2
Other	0.43	0.41	0.39	0.4	0.5	0.6
Total	15.33	14.58	13.60	11.7	12.9	15.3
Consumption						
China	2.38	2.30	2.46	2.9	3.2	4.0
Former Soviet Union	8.74	8.39	8.20	6.3	7.2	8.9
Other	2.12	2.01	1.67	1.6	1.8	2.4
Total	13.24	12.70	12.33	10.8	12.2	15.3
World Oil Consumption	66.03	66.15	66.60	70.6	76.5	85.9

legitimately puts those measures under the microscope of cost/benefit analysis. And more and more, that public is ready to question the benefits of any regulation or law that adds to the cost of a fragile economy in an increasingly competitive world, especially when other nations aren't tagging along with similar measures.

There is a general consensus today that oil and gas in their various conventional forms will continue to dominate the world's energy mix for many years to come, perhaps for most of the next century. There is little doubt that during the next 20 years or so, as more developing nations enter the once-exclusive club of industrialized nations, that demand for oil and gas will continue to grow.

Table 17-2
The Outlook–Exploration, Drilling, Production

OIL INDUSTRY OUTLOOK

THE OUTLOOK — EXPLORATION, DRILLING, PRODUCTION

	1993	1994	1995	1996	1997	1998	1999	Change 1994 to 1999	% change 1994 to 1999
DRILLING ACTIVITY U.S.									
Active drilling rigs	757	805	880	920	1,000	1,090	1,190	385	47.8
Well completions	24,936	26,520	28,670	30,170	32,660	35,540	38,610	12,090	45.6
Footage drilled (thousands)	138,775	147,500	159,690	168,350	182,570	199,020	216,600	69,100	46.8
Average well depth	5,565	5,562	5,570	5,580	5,590	5,600	5,610	48	0.9
Wells drilled per active rig	32.9	32.9	32.6	32.8	32.7	32.6	32.4	-0.5	-1.5
Footage drilled per active rig	183,322	183,230	181,466	182,989	182,570	182,587	182,017	-1,213	-0.7
REVENUE AND COST									
Average well cost ($)	377,000	385,000	398,060	412,520	428,150	444,720	459,170	74,170	19.3
Total wellhead revenues ($ million)	73,623	77,570	83,918	88,905	96,443	105,371	114,377	36,807	47.5
Total drilling expenditures ($ million)	9,401	10,210	11,412	12,446	13,983	15,805	17,729	7,518	73.6
% of wellhead revenues invested in drilling	12.8	13.2	13.6	14.0	14.5	15.0	15.5	2.3	17.8
PRODUCTION									
Capital expenditures ($ million)	2,318	2,458	2,741	3,010	3,356	3,695	4,255	1,797	73.1
Producing wells — oil & gas (End of year)	872,700	866,770	861,200	856,400	852,400	849,600	846,900	-19,870	-2.3
Abandoned wells	25,000	26,000	27,000	28,000	29,000	30,000	32,000	6,000	23.1
PRICES									
Crude oil Average U.S. wellhead ($/barrel)	14.24	15.00	16.00	15.50	16.00	17.00	18.00	3.00	20.0
Natural gas Average U.S. wellhead ($/mcf)	1.97	2.10	2.30	2.60	2.90	3.20	3.50	1.40	66.7

Even with most forecasts calling for a significant jump in oil demand in the next 20 years and a tighter supply/demand balance as capital constraints keep OPEC from adding significantly to productive capacity anytime soon (Table 17-1), the odds are that there will be no more long term oil price shocks. That's just as well for the oil and gas industry, for it assures a generally healthy out-

Table 17-3
Outlook 2003–Exploration, Drilling, Production

EXPLORATION, DRILLING, AND PRODUCTION

OUTLOOK 2003 — EXPLORATION, DRILLING, PRODUCTION

	1993	1998	2003	Change 1993 to 1998	Change 1998 to 2003	Change 1993 to 2003	% Change 1993 to 1998	% Change 1998 to 2003	% Change 1993 to 2003
DRILLING ACTIVITY U.S.									
Active drilling rigs..................	757	1,090	1,660	333	570	903	44.0	52.3	119.3
Well completions.....................	24,936	35,540	52,970	10,604	17,430	28,034	42.5	49.0	112.4
Footage drilled (thousands)	138,775	199,020	299,280	60,245	100,260	160,505	43.4	50.4	115.7
Average well depth	5,565	5,600	5,650	35	50	85	0.6	0.9	1.5
Wells drilled per active rig	32.9	32.6	31.9	-0.3	-0.7	-1.0	-1.0	-2.1	-3.1
Footage drilled per active rig	183,322	182,587	180,289	-735	-2,298	-3,033	-0.4	-1.3	-1.7
REVENUE AND COST									
Average well cost ($)...............	377,000	444,720	524,360	67,720	79,640	147,360	18.0	17.9	39.1
Total wellhead revenues...........	73,623	105,371	158,731	31,748	53,360	85,108	43.1	50.6	115.6
($ million)									
Total drilling expenditures.........	9,401	15,805	27,775	6,404	11,970	18,374	68.1	75.7	195.5
($ million)									
% of wellhead revenues									
invested in drilling	12.8	15.0	17.5	2.2	2.5	4.7	17.5	16.7	37.0
PRODUCTION									
Capital expenditures.................	2,318	3,695	6,667	1,377	2,972	4,349	59.4	80.4	187.6
($ million)									
Producing wells — oil & gas....	872,700	849,600	834,000	-23,100	-15,600	-38,700	-2.6	-1.8	-4.4
(Year end)									
Abandoned wells	25,000	30,000	40,000	5,000	10,000	15,000	20.0	33.3	60.0
PRICES									
Crude oil..................................	14.24	17.00	24.00	2.76	7.00	9.76	19.4	41.2	68.5
Average U.S. wellhead									
($/barrel)									
Natural gas	1.97	3.20	4.70	1.23	1.50	2.73	62.4	46.9	138.6
Average U.S. wellhead									
($/mcf)									

look and prospects for growth in the long term. As Tables 17-2 and 17-3 show, even in the most mature producing province, where it all started, in the U.S., the oil and gas producing industry can look forward to many years of activity to come.

Exercises

Chapter 1 Exercises

1. The Latin words _____ and _____ mean _____ and _____ and are the derivative of the word petroleum.

2. The term _____ _____ generally refers to both the oil and gas industries.

3. All petroleum deposits discovered thus far originated in the last _____ of the earth's life during the Paleozoic, Mesozoic, and Tertiary Cenozoic eras.

4. The historical development of the U.S. petroleum industry consists of _____ distinct stages.

5. A period of wild exploration began in _____ when "Colonel" Edwin L. Drake drilled the first commercially successful oil well near Titusville, Pennsylvania.

6. The third historical stage of the petroleum industry is sometimes called the _____ era of the petroleum industry.

7. Today the total proved crude-oil reserves of the Free World is approaching _____ barrels.

8. The _____ brought the Great Depression and government regulation of production.

9. The petroleum industry's chief troubles usually have arisen from _____: too much output, too little demand, too low a price, too little profit.

10. After _____, a period of competitive realignment found some companies with substantial foreign crude oil resources and others almost exclusively dependent on domestic production.

Chapter 2 Exercises

1. The six largest oil companies in the United States are _____, _____, _____, _____, _____, _____.

2. An _____ producer is a person or corporation that produces oil for the market but usually does not have a pipeline or refinery to move or process that product.

3. Sometimes these operators are true _____, leasing and drilling on small parcels of land that the majors either overlooked or didn't think worth fooling with—until a discovery is made.

4. Independents do most of their drilling _____ in the _____ states.

5. Historically, independent producers have concentrated their efforts in the _____ and _____ areas.

6. Historically, independent producers have drilled about _____ of the new field wildcat wells in the U.S.; while the majors accounted for only about _____.

7. In absolute terms, independents' production has remained more or less _____ since 1977.

8. Declines in domestic production have been borne for the most part by the _____ _____ _____.

9. Two trends sweeping the world petroleum industry structure, _____ and _____, are creating business opportunities in many countries.

10. The single aspect of a petroleum company's structure that most distinguishes its type of operation is whether or not it is _____; that is, having upstream and downstream operations.

Chapter 3 Exercises

1. The earth's crust is composed of three types of rocks: _____, _____, _____.

2. Sedimentary materials can be classified as (1) _____ (sandstone, shale), (2) _____ (certain limestones), (3) _____.

3. Although _____ rocks are associated with oil, not all _____ rocks contain oil.

4. Early life began in vast _____ and inland _____ that covered large portions of the present continents.

5. Crude oil and natural gas occur as fluids that occupy the _____ _____ of the sedimentary rocks.

6. The ratio of the pore volume to the total rock volume is known as _____, commonly expressed as a percentage.

7. The ease with which fluid moves through the interconnected pore spaces of rock is called _____.

8. Oil and gas continue to migrate, ever moving upward, until they are finally _____ by some kind of deformation in the strata or layers of rock.

9. Traps are classified into three major types: _____, _____, and _____.

10. What three things must occur for petroleum to accumulate?

Chapter 4 Exercises

1. A _____ negotiates directly with landowners or lease brokers to acquire acreage.

2. The two kinds of drilling are called _____ and _____.

3. Of the two kinds of operations—in house or contract—more than _____ of all wells in the Free World are drilled by independent contractors.

4. Over the years, three standard types of drilling contracts have evolved: (1) _____, (2) _____, and the (3) _____.

5. Even though competent geological advice may be available, the early period of development of a field is _____.

6. A very important aspect of field development is _____.

7. Usually an operator wants to develop the largest possible area with the fewest _____ of wells without running the risk of locating a well beyond the limits of the accumulation and drilling a dry hole.

8. Some _____ is usually followed in placing wells.

9. The distance between the wells drilled is known as _____.

10. When planning a development pattern, what 6 factors come into play?

Chapter Five Exercises

1. The most common type of land drilling rig is the _____ _____, sometimes called a _____ rig.

2. In shallow water or swamps, a _____ is used to drill.

3. A _____ _____ operates in water as deep as 350 ft.

4. _____ _____, yet one more of many kinds of offshore vessels, are pinned to the seafloor by long steel pilings.

5. _____ _____ can also operate in water from 200-500 meters deep and they are equally stable, but they are not fixed.

6. Whether drilling on land or offshore, a successful drilling system must provide what 3 things?

7. Today, the two most widely used drilling methods are _____ and _____ _____.

8. Drilling with a cable-tool rig is a _____ process.

9. In rotary drilling, a bit used to cut the formation is attached to steel pipe called _____.

10. The material used to lift the cuttings from the bottom of the hole and to control pressures encountered during rotary drilling is called the _____ _____.

Chapter Six Exercises

1. List 4 widely used well logs for formation evaluation.

2. Each drilling prepares a _____ _____, which is a record of operations and progress during his tour.

3. The log which shows the type of rock, specific formation penetrated, age of the rock, well depth at which formation was encountered, and indications of porosity, permeability, and oil content is called a _____ _____.

4. _____ _____ is the continual inspection of the drilling fluid and cuttings for traces of oil and gas.

5. The _____ _____ is a computer analysis of certain drilling parameters and data.

6. The _____ _____ is a record of core analysis data and lithology versus depth.

7. One of the largest categories of logs are the _____ _____ which provide data that is very important in formation evaluation.

8. The most widely used wireline log today is known as the _____ _____.

9. _____ was one of the earliest formation evaluation methods.

10. _____ _____ is a supplemental coring methods used in zones where core recovery by conventional methods is small or where cores were not obtained as drilling progressed.

Chapter Seven Exercises

1. What two types of perforating guns are commonly used?

2. A _____ well completion is a technique in which the tubing is run and the wellhead is assembled only once in the life of the well.

3. A _____ completion allows simultaneous production of two or more separate productive intervals.

4. _____ _____ completions are used in an area where unconsolidated (loose-grained) sand is encountered which may enter the wellbore, eroding equipment and plugging flow lines.

5. Generally, operators want to produce as little as possible _____.

6. Open-hole, or _____, completions are situations in which the oil string is set on top the indicated productive interval, leaving the productive interval as an open borehole with no pipe to protect it.

7. _____ completions is a term applied to a well that has been drilled and completed in some form of horizontal or near-horizontal situation.

8. In reservoirs composed of alternating layers of productive sands separated by shale or dense sections, _____ _____ and _____ often reduce or exclude water production.

9. One of the more useful survey tools for determining water entry into a wellbore is a _____ _____ _____ tool.

10. In order for a well to produce, a pressure _____ or gradient must be established between the well and its drainage radius, the area around the well that contains hydrocarbons.

Chapter Eight Exercises

1. List the 5 kinds of casing.

2. _____ _____ is the conduit that raises drilling fluid high enough above the ground level to return the fluid to the mud pit.

3. The second string of casing to be set, which protects freshwater sands from contamination by oil, gas or salt water from the deeper formations is called the _____ _____.

4. An _____ _____, though not always run, protects the hole against loss of circulation in shallow formations.

5. Unlike casing that is run from the surface to a given depth and overlaps the previous string, a _____ is run only from the bottom of the previous string to the bottom of the open hole.

6. The _____ _____ is a heavy, blunt object placed on the bottom of the casing which prevents the bottom of the casing from deforming.

7. _____ _____ are multipurpose devices that permit the casing to float into the wellbore.

8. _____ and _____ are attached to the casing to aid the cementing process.

9. The _____ is the casing attachment to the BOP or the production christmas tree.

10. The _____ casing is nearly always welded to the wellhead.

Chapter Nine Exercises

1. Oil recovery is a _____ process.

2. In a _____-_____ reservoir, oil is displaced when the gas is liberated from solution in the oil.

3. In a _____ _____ reservoir, the agent is the free gas cap originally present and overlying the oil-bearing zone.

4. In a _____ _____ reservoir, water from adjacent aquifers encroaches into the oil-bearing portion of the reservoir.

5. Gas-cap and water drives are much more efficient than dissolved-gas drives, with _____ _____ generally the most efficient.

6. In each of the recovery mechanisms, _____ exerts a supplemental effect upon the process and must be considered when vertical direction is involved.

7. The factors affecting recovery are: _____, _____, _____, _____.

8. _____ _____ depends on the degree to which the advancing gas or water invades the entire reservoir and how uniformly the gas or water displaces or flushes the oil.

9. Recovery from most pools is directly dependent on the _____ of _____.

10. The _____ _____ _____ for an oil reservoir is defined as the highest rate that can be sustained for an appreciable length of time without damaging the reservoir and which, if exceeded, would minimize ultimate oil recovery.

11. In establishing the maximum efficient rate for a reservoir, two independent _____ and one _____ must be satisfied.

12. In the early stages of field development, _____ is usually limited by the efficiency of the individual wells.

13. One of the most useful tools in determining productive capacity of a well is the _____.

Chapter Ten Exercises

1. To protect the casing, a smaller-diameter string of pipe called _____ is run into the well.

2. From the wellhead, the produced fluids are transported through a flow line to a field gathering station, usually called a _____ _____.

3. Producing wells are normally classified by the type of _____ used to get the produced fluids from the bottom of the well into the flow line.

4. There are many variations in the design of a _____ _____ system, but the basis concept is to take gas from an external source and inject it into the produced fluids within the tubing string.

5. The most well-known and widely used artificial lift technique is called _____ _____.

6. The four basic types of rod pumping units are the _____ _____, the _____, the _____, and the _____.

7. A _____ _____ _____ is a specially designed centrifugal pump with a shaft that is directly connected to an electric motor.

8. _____ _____ _____ is a method of pumping oil wells using a bottom-hole production unit consisting of a hydraulic engine and a direct couples positive displacement pump.

9. The _____ of crude oil is very important, since the selling price of crude oil is a function of its _____.

10. Increasing the operating _____ of a separator will increase the API gravity and will reduce the GOR.

Chapter Eleven Exercises

1. The _____ is the cast or forged steel configuration of pipes at the top of a wellbore that controls the well's pressure at the surface.

2. The _____ is designed to support the tubing string, to seal off pressures between the casing and tubing, and to provide connections at the surface to control liquid or gas flow.

3. High pressure wells are equipped with special heavy valves and control equipment known as a _____ _____.

4. A _____ is a piece of equipment used to separate wellstream gas from free liquids.

5. Methods used to prevent hydrates from forming in a gas line are called _____ _____ _____.

6. Crude oil is produced with various quantities of _____, _____, and _____ mixed with the oil.

7. _____ are mixtures of fluids.

8. One of the most common types of _____ _____ are heater treaters.

9. The _____ _____ is used to separate free gas and free water from free oil and emulsions.

10. The settling vessel used for separation of unstable emulsions is called a _____ or _____.

11. Once the oil is clean enough to meet pipeline specifications, it is flowed into storage tanks, sometimes called _____ _____.

12. Before a tank battery is put into service, the storage tanks are _____.

13. Measurement of oil and gas production or sales is sometimes called _____.

14. Automatic measuring devices for sale or transfer to the pipeline are called _____ units.

Chapter Twelve Exercises

1. _____ _____ is probably the most common type of production problem.

2. Special equipment called a _____ or _____ is used to remove and install subsurface equipment in a well.

3. In wells which produce from loosely consolidated sandstone formations, a certain amount of _____ is usually produced with oil.

4. _____ _____ is a common problem which occurs when something happens to the formation near the wellbore, slowing oil production.

5. _____ accumulation in the tubing and at the surface flow lines is a problem in some areas in which a special type of crude oil known as "paraffinic crude" is produced.

6. _____ of oil and water are a fourth kind of common production problem.

7. _____ of equipment is one of the most costly problems plaguing the oil industry.

8. Disposing of _____ produced with oil can be expensive.

9. _____ operations are major remedial operations sometimes required to maintain maximum oil producing rates.

10 Before beginning work on a well, it is first necessary to _____ the well with some fluid, such as drilling mud, salt water, oil or possibly a special workover fluid, which has sufficient hydrostatic pressure to counteract the formation pressure when the hole is filled with the fluid.

Chapter Thirteen Exercises

1. When tests indicate that a well may be an economical producer but for some reason the rate of flow is inadequate, the formation may be _____ to increase the well's productivity.

2. _____ was first used for well stimulation in 1895.

3. _____ _____ is the most commonly used for acidizing treatments because it is economical and leaves no insoluble reaction product.

4. In the conventional controlled _____ _____, tubing must be in the hole and the well must be capable of being filled with fluid.

5. Another type of controlled treatment is the _____ _____.

6. The advantage of the packer method is that the acid is confined to the section of the formation below the _____.

7. Overall, the advantage of _____ _____ is that maximum benefit is derived from the acid by controlling the section into which it enters.

8. _____ _____ is for treating dense and tight limestones.

9. Sometimes in limestone formations, when there is a possibility of breaking into a water zone, _____ _____ are used.

10. Another application of _____ _____ is to clean up the contaminated zone near the wellbore with one stage of acid.

11. _____ are dissolved in acid solutions to slow the reaction rate of the acid with metals.

12. Intensified acid is a mixture of inhibited hydrochloric and _____ acids.

13. _____ are chemical additives that lower the surface tension of a solution.

14. _____, when added to an acid solution, are chemical agents that counteract natural emulsifiers in crude oil.

15. _____ _____ removes the mud cake from the face of the productive interval during completion and before subsequent workover jobs.

16. In _____ _____, developed about 1948, oil or water, mixed with sand or other propping material, is pumped into the formation at a high rate, causing fractures in the formation.

17. _____ fracturing fluids are a mixture of water and acid.

18. The most common propping material in the U.S. is _____ _____.

Chapter Fourteen Exercises

1. The recovery of additional oil after primary methods may occur with the operator applying _____ _____ _____ _____.

2. EOR methods are usually divided into _____, or waterflooding, and _____, commonly called EOR.

3. In waterflooding, water is injected into a reservoir for _____.

4. The _____ and _____ of a reservoir control the location of wells and, to a large extent, dictate the methods by which a reservoir may be produced.

5. Most water injection operations are conducted in fields with only _____ structural relief where the oil accumulates in stratigraphic traps.

6. The total recovery of oil from a reservoir is a direct function of _____, since _____ determines the amount of oil present for any given percent of oil saturation.

7. One of the major considerations in waterflooding is where to locate enough _____ for the operation.

8. Operators can choose to use _____ or _____, depending on the operation and the reservoir.

9. Four basic patterns are usually chosen: (1) _____, (2) _____, (3) _____, and (4) _____.

10. Occasionally _____ _____ are conducted to evaluate the flood potential for a new project.

11. If an operator determines that there is an appreciable amount of producible oil remaining in the reservoir, he may opt for a third stage of production: _____, or _____.

12. _____ _____ include polymer flooding, surfactant flooding, and alkaline flooding processes.

13. _____ _____ usually use carbon dioxide, nitrogen, or hydrocarbons as solvents to increase oil production.

14. _____ _____ processes add heat to the reservoir to reduce oil viscosity and/or to vaporize the oil.

15. _____ _____ generally occurs in two steps: steam is pumped down a producing well to heat and thereby loosen oil near the producing wellbore, then steam is driven down an injection well and moves through the reservoir toward a producing well, pushing the warm, moveable oil in front of it.

Chapter Fifteen Exercises

1. Natural gas processing and cycling plants recover _____ _____ from gaseous streams produced directly from gas wells or from normal oil and gas separation equipment on oil wells.

2. _____ or _____ gas usually means a stream that is potentially worth processing for its liquid.

3. The term _____ _____ usually refers to gas obtained from a well in association with crude oil.

4. _____ _____ is any gas suitable for sale as commercial natural gas that comes from a processing plant.

5. _____ _____ means that the content of hydrogen sulfide, other sulfur compounds, and carbon dioxide is low enough that the gas may be sold commercially without further effort to remove these compounds.

6. Natural gas liquids include all hydrocarbons that have the properties of the _____ _____ ethane through decane with no more than trace amounts of anything heavier.

7. _____ _____ is a hydrocarbon product composed predominately of propane and/or propylene and have a vapor pressure of not more than 215 psig at 100∞F.

8. _____ _____ is a hydrocarbon product containing predominately butanes and/or butylenes and have a vapor pressure of not more than 70 psig at 100∞F.

9. _____ _____ is a product extracted from natural gas and is used as refining blending stock.

10. _____ is in growing demand as a base for the manufacture of plastics, alcohols, and other chemicals.

11. _____-produces two forms of energy such as heat and power from a single fuel which offers the petroleum industry a major new growth area. Whether to improve gas demand or process plant economics, or an investment in a larger role as an electrical power producer.

12. Cogeneration power sales help offset _____ costs, enabling the operators to replace crude with gas to generate steam.

13. Cogeneration will benefit the utilities over the long run because it will recapture the _____ power generating market they have lost while avoiding the cost of replacing or building new base load plants.

14. For the near term, the only significant growth in power generating capacity will be in _____.

Chapter 16 Exercises

1. A _____ well exposes more of a formation to the wellbore than does a _____ well.

2. Name the four types of horizontal well radiuses.

3. A horizontal well can produce _____ as much oil as a vertical well in 5 years.

4. Name some kinds of formations that can benefit from a horizontal well.

5. _____ uses special electronic tools to impart well status data from down-hole to surface during drilling.

6. A _____ _____ unit injects and retrieves a continuous string of tubing into a larger string of tubing or casing.

7. What is the primary advantage of coiled tubing in producing operations?

8. Name at least five coiled tubing applications.

9. What is the most significant advance in seismic technology of the past 20 years?

10. 3-D seismic is now seen as a key tool for delineating _____ _____ and analyzing _____ _____.

Chapter 17 Exercises

1. For as long as the industry has existing, there have been warnings about its _____ _____.

2. Scarcity of petroleum inevitably leads to _____ _____, which in turn leads to the incentive to find more petroleum.

3. World crude oil reserves today total almost _____ _____ bbl.

4. True or false: Oil reserves have fallen sharply since the 1986 price collapse.

5. The latest prophecy of doom for the petroleum industry comes from
   _____ _____.

6. The 1990s became know as the Decade of _____.

7. New federal environmental requirements being considered could cost
   _____ jobs in E&P directly and _____ jobs indirectly.

8. Most forecasts call for a significant (drop, leveling, or increase?) in oil
   demand the next 20 years.

Glossary

A

acoustic log/logging   recorded measurement of ultrasonic signal travel through formation rock in order to identify formation rock lithology, porosity, and fluid saturation.

acid intensifier   intensifying additive that acts to accelerate or strengthen chemical reaction of the acid.

acid inhibitor   inhibiting additive that acts to top or retard chemical reaction of the acid.

anticline   a convex geological strata in the form of an elongated dome in which petroleum may accumulate.

annulus   downhole space between drillstring or casing and borehole wall, or between production tubing and casing, or between surface casing and production tubing and casing.

API gravity   gravity (weight per unit of volume) of crude oil or other liquid hydrocarbon as specified by the American Petroleum Institute.

artificial lift   mechanical aid to production when natural drive is insufficient.

asphaltene   any of the dark, solid constituents of crude oils and other bitumens.

automatic tank gauge   device that indicates the height of a liquid in a tank according to the position of a float on one end of a line.

B

bailer a bucket-shaped cylinder used in cable tool drilling to remove rock cuttings and mud from the borehole.

barite weighting material (mineral) used to increase density of drilling fluid.

barge a non-self-propelled marine vessel used to carry, support, and store materials or units in offshore operations.

basic sediment and water (BS&W) impurities found in crude oil.
bedding plane a division plane which separates individual strata or beds in rock.

bentonite highly absorbent rock composed principally of clay minerals and silica.

blowout uncontrolled gas and/or oil pressure and eruption from a well.

blowout preventer a safety device installed immediately above the casing or in the drillstem that can close the borehole in an emergency.

bumping the plug increase in cementing pump pressure noted when the top plug reaches/bumps the float collar.

C

cable tool drilling making a hole in the earth by repeatedly slamming a tool suspended by a cable into the earth.

caliper log/logging record of hole size diameter vs. depth.
cantilevered mast/jackknife derrick rig a derrick rig whose legs are hinged at the base so that the rig can be lowered and transported intact.

casing steel pipe used in wells to seal the borehole from formation fluids and to reinforce the walls of the borehole.

casing completion completing a well by installing casing within the borehole.

casing hanger device used to support or suspend the casing within the well bore. Usually consisting of mechanical teeth to keep the pipe from slipping.

casinghead gas gas produced with oil from an oil well.

cement accelerator cement additive that promotes cement setting and reduces waiting time.

cement additive additive ingredient to cement to achieve special purpose or function, e.g., weight control, fluid-loss reduction, setting time control.

cement bond log/logging record of cement bond or bonding detected along well casing after cementing.

cement retarder cement additive that delays cement setting and promotes cement pump-ability in deep, hot wells.

cementing securing the well casing in place and excluding excess fluids from the wellbore by forcing cement slurry down through the casing and out up the annulus.

check valve valve which operates to allow flow in only one direction.

chelating agent organic compound in which atoms form multiple coordinate bonds with metals in solution.

christmas tree casinghead valve assembly through which a well is produced.

cogeneration simultaneous generation of electricity and process steam or heat.

collar locator log/logging record of depth and position of drillstem collars.

complexing agent substance used to form a complex compound with another material in solution.

conductivity property of, capacity for, or tendency toward conductance of an electric current, varying according to formation composition.

conductor pipe large diameter casing used to keep the top of the wellbore open and to convey upflowing drilling fluid from the borehole to the mud pit.

connate water water in the he same zone of a petroleum reservoir occupied by oil or gas.

controlled acidizing treatment pumping acid solution down a well into the producing zone, controlled by minimal displacement fluid in the tubing and wellbore.

conventional perforated casing completion completing a well by installing and perforating casing within the borehole.

core analysis analysis of rock samples which are cut downhole and brought to the surface for examination.

core barrel a specialized tube with cutting edges on the bottom which is used to obtain core samples from the bottom of the well.
core log record of core analysis data and lithology vs. depth.

coring cutting rock samples downhole and bringing them to the surface for examination.

coring bit a specialized drilling bit for cutting and removing rock samples from the bottom of a well.

corrosion chemical or electrochemical deterioration of metal.

corrosion inhibitor additive or agent or treatment used to stop or inhibit corrosion.

crude oil unrefined petroleum, i.e., oil as it comes from the well.

cuttings rock chips and fragments resulting from drilling that are brought to the surface in the circulating drilling mud.

D

day/day-rate contract a well drilling contract which specifies paying the drilling contractor on the basis of days required to achieve the well planned.

demulsifier/demulsifying chemical/chemical action used to break down crude oil/water emulsions by reducing surface tension of the oil film surrounding water droplets.

derrick structure erected over a well site to support drilling equipment and a mast for raising and lowering drillpipe and casing.

development drilling drilling wells in a resource area already proven to be pro-ductive.

diamond bit drilling bit with industrial diamonds set in cutting surfaces.

dip log/logging record of formation dip vs. depth.

directional drilling non-vertical well drilling.

directional log/logging record of borehole deviation.

dissolved gas drive See solution gas drive/dissolved gas drive.

dome incursion of an underground formation into a formation above it, sometimes piercing the overlying formation.

drag/fish tail bit drilling bit with cutting teeth/tooth in the shape of a fish tail, drilling being accomplished by tearing and gouging, especially in soft forma-tions.

drainhole completion well completion in some form of horizontal or non-ver-tical hole configuration.

drawworks collective term applied to hoisting, clutching, power, braking, and other machinery used on a drilling rig.

drill collar heavy tubular connector between drillpipe and drilling bit.

drillstem drillpipe. See also drill string.

driller's log record of the driller concerning well spudding, hole size, bits use, tool use depth, footage, casing setting, and unusual conditions downhole. driller's time log record of the time required for the drilling bit to drill rock of specific interest.

drill string the string of tools that are used to drill a well, i.e., the kelly, drillpipe, drill collars, stabilizers, and drilling bit.

drilling bit/drill bit a cutting or grinding tool or head attached to the tip of the drill string.

drilling break change in penetration rate during drilling.

drillpipe pipe used in making up the drillstem, an essential part of the drill-string.

drillship/drilling ship a ship outfitted with a drilling derrick.

drillstem testing obtaining fluid sample from a formation using a tool attached to the drillstem.

dry hole a well drilled unsuccessful, i.e., petroleum was not found.

dry/lean gas natural gas containing insignificant amounts of associated petroleum liquids.

E

electric log/logging measurement of electrical properties of formations and formation fluids.

emulsion drilling fluid/mud drilling fluid formulated with an oil emulsion base.

enhanced oil recovery usually refers to tertiary recovery methods which alter oil properties in the reservoir in order to improve recovery.

F

fault fracture of earth's crust accompanied by the shifting of one side of the fracture with respect to the other side.

fingering infiltration of water or gas into an oil-bearing formation during production.

fixed-platform rig a rig fixed to the sea floor by pilings.

flow test determination of productivity of a well by measuring total pressure drop and pressure drop per unit of formation section open to a well during flow at a given production rate.

fluid-loss additive cement additive that reduces filtration rates to prevent fluid loss during cementing operations.

foam/mist drilling drilling using low-density fluid in which bubbles are purposely entrained.

fold/folding bend, curve, or deformation of layered rock without breaking or faulting.

footage/footage-rate contract a well drilling contract which specifies paying the drilling contractor on the basis of footage as the well is drilled.

formation damage condition in the formation near the wellbore that inhibits production.

formation evaluation appraising the hydrocarbon potential and content of a geologic formation using techniques including drilling fluid and cuttings analysis logging, coring and core analysis, wireline logging, sidewall coring, wireline formation testing, and drillstem testing.

frac/fracturing treatment formation treatment whereby cracks are initiated in the rock in order to improve permeability.

free gas natural gas in a free state contained in reservoir rock above a more dense reservoir fluid, e.g., crude oil.

friction loss mechanical energy loss because of mechanical friction between moving parts.

friction-reducing additive cement additive that promotes turbulent flow of cement al low displacement rates.

G

gamma ray log/logging recorded measurement of natural formation radiation in order to identify formation rocks.

gas cap compressed natural gas in a free state contained in reservoir rock above a more dense reservoir fluid, e.g., crude oil.

gas cap drive reservoir drive provided by the expansion of compressed gas in a free state above the reservoir fluid being produced.

gas conditioning natural gas treatments including oil/gas separation, emulsion treatment, and cleaning and decontamination.
gas lift use of compressed gas to decrease density and help lift oil from the bottom of a well to the surface.

gas life mandrel tool or device designed to hold the gas lift valve assembly; may be the conventional type, the wire-line eccentric type, the wire-line concentric type, or the capsule type, in which the mandrel completely surrounds the valve. They are adapted to provide either tubing or casing flow.

gas-oil ratio measurement of the volume of gas produced with oil.

gas processing natural gas operations whose primary purposes are the recovery of gas liquids.

gauging measuring

gel viscous substance which can form an emulsion with water or oil for use in suspending sand or other proppants.

gravel packing filling the cavity around the borehole with gravel to prevent hole caving and/or sand production.

gun barrel wash tank or settling vessel in which an oil and water emulsion can separate.

H
heavy oil crude oil having 20 SYMBOL 176 \f "Symbol"API gravity or less.

heavyweight cement additive cement additive that increases cement slurry density.

horizontal integration business operations in similar fields or supporting operations.

huff and puff cyclic steam/hot water injection into a well in order to stimulate oil production.

hydrate/gas hydrate icy lattice containing gas molecules, causing blockage of flowlines and pipelines transporting natural gas.

I

independent producer a producer of petroleum without vertical integration of operations.

interfacial tension surface tension occurring at the interface of two liquids.

J

jackknife derrick rig *See* cantilevered mast/jackknife derrick rig.

jackup rig a rig with retractable legs that can be raised or lowered in order to raise or lower the drilling platform.

joint fracture or split in rock surface without noticeable displacement of movement of rock.

K

kelly square or hexagonal hollow shaft of which one end engages the drilling table and the other end engages the drillpipe.

kick pressure surge in the well.

L

lag time the amount of time it takes for cuttings to circulate from the bottom of the hole to the surface.

landman manager of landower relations, including securing of leases, lease amendments, and other necessary agreements.

laterolog logging instrument in which electric current is forced to flow radially through the formation.

lightweight cement additive cement additive that reduces cement slurry density.

liquefied natural gas gas (mainly methane) that has been liquefied in a refrigeration and pressure process.

liquefied petroleum gas lighter hydrocarbons (mainly propane and butane) that have been liquefied by special processes.

lithology physical characterization of rock based on color, structure, mineralogy, and grain size.

lost circulation loss of substantial quantities of drilling mud into a formation that is pierced by the drilling bit.

lost circulation additive cement additive that prevents lost circulation during cementing operations.

M

major/major producer the largest United States oil companies, e.g., Exxon, Mobil, Texaco, Chevron, and Amoco.

making a trip See tripping/pipe tripping.

mast derrick held upright by guy wires, a frame to suspend components used to drill wells or for workover, or a gin pole.

microlaterolog a resistivity logging instrument with one center electrode and three circular ring electrodes around the center electrode.

microlog a resistivity logging instrument with electrodes mounted at short spacing in an insulating pad.

migration movement of hydrocarbons in the ground; primary migration is from source bed or rock to permeable rock.

mineral right ownership of minerals that may exist within a property tract.

mobility ratio ratio of mobility of a driving fluid to that of a driven fluid.

mud cake build-up of drilling fluid and cuttings particles that coats the wall of the borehole.

mud log/mud logging analysis of cuttings brought to the surface in circulating drilling fluid.

mud system equipment comprising drilling fluid storage and circulating system components.

multiple-zone completion completing a well in which provision is made for producing formation fluids from more than one formation or formation zone.

multistage cementing cementing of multiple formation intervals behind the well casing, using ported couplings.

N

natural gas the gaseous forms of petroleum such as mixtures of hydrocarbon gases and vapors.

natural gasoline light liquid hydrocarbon mixture recovered from natural gas or present at the wellhead as a condensate.

neutron log/logging recorded measurement of artificially produced radiation within a well in order to measure formation fluids.

O

oil-base drilling fluid/mud drilling fluid formulated with an oil base.

oil/gas show evidence of oil and gas contained in cuttings or circulated drilling fluid.

oil string the length of well casing through which oil is produced from the formation.

open-hole/barefoot completion well completion method in which the oil string is set above the productive interval, leaving the latter as an open borehole.

open-hole log log such as an electric log that is run only in uncased wells.

overburden earth material overlying a mineral or other useful material deposit.

P

packer expanding plug used to seal off tubing or casing sections when cementing or acidizing or when isolating a formation section.

packer method controlled acidizing treatment in which the packer prevents acid from traveling further up the well annulus than necessary.

pay section producing formation.

perforate/perforating making holes through the casing opposite producing formation to allow flow into the well.

perforating gun/tool tool used to make holes through installed casing.

perforation depth log/logging record of perforation depth detected along well casing after perforating.

permanent well completion completing a well in which the tubing is run and the wellhead is assembled only once in the life of the well.

permeability connectedness of rock pores, a measure of rock resistance to fluid movement through it.

petroleum generic name of hydrocarbons, including crude oil, natural gas, natural gas liquids, and their refined products.

pilot flood a pilot-scale waterflooding project conducted in order to evaluate operation procedures and to assess and predict performance of the waterflood.

plugback sealing well casing to separate producing intervals in the wellbore from depleted intervals.

pore/porosity a void or empty space in rock; a property of petroleum-bearing rock.

proppant/propping agent granular substance used to keep earth fractures open when fracturing fluid withdraws.

proved reserve petroleum that has been discovered and determined to be recoverable, but is not yet produced.

R

radioactive log measurement of natural and artificially produced radiation within a well in order to identify formation rocks and fluids.

radioactive tracer log/logging record of travel of a radioactive tracer substance within a formation or within the borehole.

reading box glass-windowed box on the outside of a tank in which a line from an automatic tank gauge is suspended to show the height of liquid in the tank.

reid vapor pressure a measure of the vapor pressure in pounds pressure of a sample of gasoline at 100 SYMBOL 176 \f "Symbol"F.

reservoir drive energy or force in a reservoir that causes reservoir fluid to move toward a well and up to the surface.

reservoir rock sedimentary rock formations that contain quantities of petroleum.

residue gas gas remaining after natural gas is processed and liquids are removed.

resistivity property of, capacity for, or tendency toward resistance of passage of an electric current, varying according to formation composition.

retarded acid acidizing solution whose reactivity is slowed so that the acid can penetrate deeper into a formation before being spent.

rolling-cutter/roller bit rock-cutting tool comprising shanks welded together to form a tapered body; each shank supports a rotating cone-shaped wheel with cutting teeth.

rotary drilling making a hole in the earth by rotating an entire drillstring into the earth; the tip of the drillstring may be any kind of drilling bit.

rotary table turning section of the derrick floor, transfers engine power into turning motion to rotate drillstring.

round trip See Tripping/pipe tripping.

royalty a share of petroleum production from a property without bearing the expense of finding or producing the petroleum.

S

salt-saturated cement cement formulation designed for use in salt formations and in shale formations that are water sensitive.

sample log record of rock cuttings sampled while a well is being drilled with respect to depth.

sand-exclusion completion completing a well in which sand production is discouraged with the installation of slotted or screen well liners or by gravel packing the borehole.

sand production producing sand in well fluid.

screen-out well congested with sand which has fallen out of produced fluids in the well.

secondary recovery usually refers to waterflooding treatment of the formation in order to drive reservoir fluids to the wellbore.

sedimentary rock rock formed by successive layers of materials, which are compacted by layers of subsequent deposits.

self-potential/spontaneous potential amount of electrical voltage exhibited by a natural material.

semisubmersible rig a floating drilling platform supported by underwater pontoons.

sequestering agent substance that removes metal ion from a solution system (can be either chelating or complexing agent).

service/pulling unit specialized equipment used in workover operations to recover equipment in the hole and to return the well to production.

shoe a protective plate.

sidewall coring cutting rock samples from the wall of the borehole and bringing them to the surface for examination.

single-stage cementing cementing of a single formation interval behind the well casing.

sink pressure gradient surrounding the borehole within its drainage radius.

slug a measured amount of liquid used to displace or force fluid flow in the reservoir.

solution gas natural gas in solution in reservoir fluid, e.g., crude oil.

solution gas drive/dissolved gas drive reservoir drive provided by the expansion of natural gas in solution in reservoir oil.

sour gas natural gas containing relatively large amounts of sulfur/sulfur compounds.

source bed/source rock original site of deposition of petroleum, not always the site of present accumulation.

spent acid acidizing solution sludge or residue.

squeeze cementing forcing cement into perforations, cracks, and other openings in the wall of a borehole.

stage separation use of multiple gas/oil separators in series to change API gravity of oil and gas/oil ratio.

steam soak steam/hot water injection into a well in order to stimulate oil production.

step-out well a well drilled in an unproven area that is adjacent to a proved producing area.

stock tank storage tank for oil on the lease.

strata layers of sediments or sedimentary rock.

stripper/stripper well an oil well that produces a limited amount of oil, e.g., no more than 10 bbl/day.

sucker rod a series of steel rods connecting the pump inside well tubing downhole to the pumping jack on the surface.

surface equipment lease equipment used to produce hydrocarbons and clean them in the field.

surface right ownership of a property tract with or without concurrent ownership of the mineral right for that tract.

surfactant/surface active agent chemical additives that lower the surface tension of a solution.

swabbing cleaning out a well with a special tool connected to a wireline.

sweet gas natural gas containing little sulfur/sulfur compounds.

syncline a concave geological structure in which folded strata may form traps.

T

tank battery field equipment comprising oil and gas separating components, treating equipment, and storage facilities.

tank strapping measuring tanks and computing the volume that can be contained in each interval of a tank.

tank table tabular report of the capacity of a tank computed according to the height of liquid contained in a tank.

temperature log/logging record of wellbore temperature vs. depth.

thermal recovery tertiary recovery methods in which the oil the reservoir is affected by thermal treatments, including in situ combustion, steam injection, and other methods of reservoir heating.

tertiary recovery enhanced oil recovery methods in which the properties of the oil in the reservoir are altered in order to improve recovery.

trap a geological structure that inhibits free migration of petroleum and concentrates the petroleum in a limited space; structural traps confine petroleum as a result of a structural condition in reservoir rock (fold or fault); stratigraphic traps confine petroleum as a result of variations in rock lithology (rock type or porosity); combination traps confine petroleum as a result of both structural and stratigraphic features.

tripping/pipe tripping exiting (tripping out) or entering (tripping in) the well bore with the drillstring, or both, in order to change a drilling bit. Tripping out and in is "round tripping."

tour work shift period of a member of the drilling crew.

tubing small-diameter pipe used as a flowline within a well.

turnkey contract a well drilling contract which specifies paying the drilling contractor on the basis of completing drilling and preparing a well for production.

U
unconformity surface separating two dissimilar rock formations.

uncontrolled acidizing treatment pumping acid solution down a well, followed by displacement fluid to force the acid out into the formation.

V
vertical integration business operations designed to participate in successive operations of an industry, e.g., an oil company that produces, refines transports, and markets products.

viscosity the ability of a liquid to flow.

W
wash tank a settling vessel in which an oil and water emulsion can separate.

water-base drilling fluid/mud drilling fluid formulated with a water base.

water drive   reservoir drive provided by the force of water under pressure below the reservoir fluid being produced.

waterflooding   injection of water with or without additives in order to drive reservoir fluids to the wellbore and improve recovery.

water/gas exclusion completion   completing a well in which water and/or gas production is discouraged by means of squeeze cementing and perforating.

weighting material   substances such as barite which are added to drilling fluid to increase its unit weight per gallon.

well log   geological, formation attribute, and hydrocarbon potential data obtained with special tools and techniques used downhole.

well spacing   the geographic placement of wells according to regulatory requirements and/or reservoir engineering recommendations.

well stimulation   treatments on a well or formation to obtain improved production, e.g., hydraulic fracturing, acidizing, or sand control.

wellhead   casing attachment to the blowout preventer or the production christmas tree, bolted or welded to conductor pipe or surface casing.

wet/rich gas   natural gas containing significantly large amounts of associated petroleum liquids.

wettability   ability of a solid surface to be wetted when contacted with a liquid.

wildcat well   an exploratory well drilled in an unproven resource area.

wiper plug   oil well tubing plug with capability of wiping interior of pipe during installation.

wireline log   data descriptive of downhole drilling conditions or subsurface features obtained by the use of tools, devices, and instruments lowered into the borehole on a wireline cable.

wireline logging   logging or measuring downhole formation attributes using special tools or equipment lowered into the borehole on a wireline (slender steel cable).

Index